Monika Kirloskar-Steinbach,
Mădălina Diaconu (eds.)

Environmental Ethics:
Cross-cultural Explorations

VERLAG KARL ALBER

Monika Kirloskar-Steinbach,
Mădălina Diaconu (eds.)

Environmental Ethics: Cross-cultural Explorations

Verlag Karl Alber Freiburg / München

Monika Kirloskar-Steinbach, Mădălina Diaconu (eds.)

Environmental Ethics: Cross-cultural Explorations

Environmental Ethics: Cross-cultural Explorations offers insights into the significance of cross-cultural inquiry. The volume's rich explorations illustrate how the hitherto narrow, subjective character of value judgments in environmental ethics and aesthetics can be rectified and extended by drawing on non-Euroamerican philosophical positions. This modification could, in turn, abet the establishment of those norms that facilitate a sustained protection of endangered environments on a global scale.

The Editors:

Monika Kirloskar-Steinbach is professor of philosophy at University Konstanz, Germany. She is founding editor of the international, peer-reviewed *Journal of World Philosophies* which is currently published by Indiana University Press (USA) as an open-access journal. The journal was run by Verlag Karl Alber from 2014-2016 under the title *Confluence: Online Journal of World Philosophies*. Kirloskar-Steinbach serves as editor-in-chief of the new, primer series *Bloomsbury Introductions to World Philosophies* too. She works on world philosophies, political philosophy and social philosophy.

Mădălina Diaconu is *Dozentin* for philosophy at the University of Vienna, Austria. She is editor-in-chief of *polylog. Zeitschrift für interkulturelles Philosophieren* (Vienna) and member of the editorial boards of *Contemporary Aesthetics* (Castine, US) and *Studia Phaenomenologica* (Bucharest). She has authored nine monographs, more than 150 papers and (co)edited fourteen volumes on phenomenology, the aesthetics of touch, smell and taste, urban sensescapes, sensory design and philosophy of animality.

Monika Kirloskar-Steinbach, Mădălina Diaconu (Hg.)

Umweltethik: Interkulturelle Erkundungen

Umweltethik: Interkulturelle Erkundungen bietet Einblicke in die Bedeutung interkultureller Perspektiven. Die umfangreichen Untersuchungen des Bandes veranschaulichen, wie der bislang enge, subjektive Charakter von Werturteilen in der Umweltethik und -ästhetik durch u. a. die Bezugnahme auf nicht-euroamerikanische philosophische Positionen korrigiert bzw. erweitert werden kann. Diese Änderung könnte wiederum dazu führen, dass Normen festgelegt werden, die einen nachhaltigen Schutz gefährdeter Umgebungen auf globaler Ebene ermöglichen.

Die Herausgeberinnen:

Monika Kirloskar-Steinbach ist außerplanmäßige Professorin an der Universität Konstanz. Sie ist Gründungsherausgeberin der internationalen, referierten Zeitschrift *Journal of World Philosophies*, die bei Indiana University Press (USA) erscheint. Unter dem Titel *Confluence: Online Journal of World Philosophies* erschien diese Zeitschrift von 2014-2016 beim Verlag Karl Alber. Kirloskar-Steinbach ist auch Herausgeberin der neuen Lehrbuchserie *Bloomsbury Introductions to World Philosophies.* Sie forscht zu Weltphilosophien, Politischer Philosophie und Sozialphilosophie.

Mădălina Diaconu ist Dozentin für Philosophie an der Universität Wien und Lektorin am Institut für Romanistik der Universität Wien. Doktoratsstudien in Philosophie sowohl an der Universität Bukarest als auch an Universität Wien, Habilitation für das Gesamtfach Philosophie an der Universität Wien. Chefredakteurin von *polylog. Zeitschrift für interkulturelles Philosophieren* und Redaktionsmitglied von *Studia Phaenomenologica* (Bukarest) und *Contemporary Aesthetics* (Castine, US). Forschungsprojekte über Ästhetik und urbane Sinneslandschaften. Neun selbstständige Monographien und (Mit-) Herausgeberin von neun Sammelbänden. Publikationen zu Rumänien und Osteuropa.

Gedruckt mit Unterstützung des
Publikationsfonds der Universität Konstanz

FSC
www.fsc.org
MIX
Papier aus verantwortungsvollen Quellen
FSC® C083411

Originalausgabe

www.verlag-alber.de

Satz: SatzWeise, Bad Wünnenberg
Herstellung: CPI books GmbH, Leck

Printed in Germany

ISBN 978-3-495-49109-6

Content

Cross-cultural Explorations

Socio-political Explorations

Environmental Ethics and Cross-Cultural Explorations: An Introduction

Recent burgeoning debates on environmental ethics and aesthetics in the German-speaking countries look back on a long and prestigious tradition within German philosophy on the philosophy of nature. Remarkably, a sustained engagement with the intercultural aspects of these debates is conspicuous by its absence. In a first attempt at bridging this gap, the *Viennese Society for Intercultural Philosophy* and the *German Society of Intercultural Philosophy* organised an international workshop in Vienna in February 2018. Some of the articles in this volume were workshop presentations; others have been solicited by the editors for this collection with a special purpose: we intend to initiate a more sustained debate about issues of cross-cultural significance, especially in German academic philosophy. Given the scope of this present book, we can, however, only restrict our focus to environmental ethics.

Cosmological reflections are an inherent part of philosophical ruminations in different socio-material settings. In academic philosophy, however, environmental philosophy—and in particular environmental ethics—could establish itself as a discipline only in the last half century. North American, German and Scandinavian scholars, who were the initial pioneers in the field, drew on conceptual sources from Europe and North America in ameliorating the effects of what they perceived to be a grave environmental crisis. However, given the ubiquity of the environment in our lives, there is no plausible reason as to why this status quo has to be maintained.

Environmental ethicists began their critique of the intensive exploitation of nature by focusing on the short-sighted instrumental rationality, which drove this exploitation. In this self-critique of Euroamerican culture and modernity, they often tended to trace a simplistic linear relation between worldviews and economic developments, attributing the contemporary environmental crisis to Judeo-Christian thinking and Euroamerican metaphysics inspired by it. In their analysis, several Euroamerican societies underwent a Weberian »disenchantment« of the world. As a result, anthropocentric worldviews took the place of theocentric ones. However, the former worsened the destruction of the environment in the name of humanism, civilization and technological progress.

Some of these early proponents of environmental ethics linked up the environmental crisis to a methodological individualism propounded in the humanities too. In their view, many of the Euroamerican philosophers, who are traded as belonging the core of the philosophical canon, promoted a conflictual relationship to nature. While these figures indeed allowed for humans to possess rational faculties which would (potentially at least) bridge the ontological gap between themselves and nature, environmental ethicists saw a direct link between this conflictual relationship to nature and the relative neglect of environmental concerns. They pointed to how individual human subjects were presumed to be the sole adequate objects of philosophical analysis. Their interlinkage between each other and to the non-human environment were simply deemed to be impertinent for scholarly analysis. One result of such a methodological individualism was that the environment itself faded from view; as a result, the analysis was not geared to register the effects of a synthetically-altered environment on human life. Environmental destruction continued unabated. Developments like deforestation, pollution, the negative effects of large-scale industrial production etc. forced environmental ethicists to rethink this standard model. They increasingly began to realise that faith in one's own technological prowess was not only naïvely self-congratulatory; it was also methodologically erroneous.

It was, thus, only a matter of time before a search for conceptual alternatives began, beyond the modern equation between knowledge and power, as well as the biblical legitimation to extend human domination over all other species. Specialists started to turn towards Hinduism, Buddhism, Confucianism, Daoism, etc. in an attempt at uncovering useful resources through comparisons. However, these specialists were suspected of dabbling with ›mystical‹ sources that were beyond the ken of academic philosophy. Eugene Hargrove (1989) notoriously argued against the ›intrusion‹ of such influences, endorsing a return to a more balanced view of Euroamerican values. Positions like those of Hargrove, albeit being pioneers in the field, soon faced criticism too. In their correction of such simplistic views of world philosophical traditions, their critics underscored the need for solid analyses of these traditions.

In particular, the »comparative enterprise of external appropriation« met with heavy opposition, at least of some colleagues specialising in cross-cultural studies (Larson 1987: 153). Gerald J. Larson drew attention to the historical continuity of comparative endeavours since colonialism: »The needed ›ideas‹ for environmental ethics are presumably in short ›supply‹ in our own environment, but we recognise that there is an increasing ›demand‹ for some new intellectual commodities«. Pushed by the global demand of our own work, we tend to construe non-Euroamerican ideas and concepts as »›things‹ or ›entities‹ that can be disembedded from their appropriate frameworks and then processed and made to fit into our own frameworks«. Such a method, he concluded, is »one-dimensional, overly selective, forced, anachronistic, sociologically unsophisticated, and, perhaps, worst of all, unpersuasive« (Larson 1987: 151, 153).

Although research in the ensuing decades has not been sufficiently attentive to Larson's concern, it has in other aspects indeed moved away from its initial essentialising tendencies of segregating the largely »irrational«, »emotional«, »mythical« and »natural »East« from its very opposite, the »rational«, »cultivated« »West«. In this regard, another significant development has to be noted too: Easy compare-and-contrast exercises with Euroamerican positions are being increasingly abandoned; the latter is not automatically deployed as the default lens through which such a study should be conducted.

Since 1939, when the first East-West Philosophers Conference was held in Honolulu Hawai'i, and till today, conferences and publications contributed to the emergence of so-called »comparative environmental philosophy« (Callicott 2014: 377). A notable contributor to the debate, J. Baird Callicott, strove to establish environmental ethics as its own separate discipline, and not as a derivative branch of applied ethics. For this purpose, he sought to develop a »non-anthropocentric axiology«, with which an »incipient paradigm shift in moral philosophy« could be initiated (Callicott 1984: 299–300). Callicott not only defended Aldo Leopold against John Passmore, who had disqualified the land ethic as a supposed regression to a less-sophisticated moral view, he also authored and (co-)edited a series of publications on non-Euroamerican—mainly Asian—contributions to environmental philosophy (Callicott and Ames 1989; Callicott and McRae 2014). As a result, he outlined a spectrum of alternative concepts of the self, ranging from interpretations of Euroamerican Rational Individualism to Hindu Universal Essentialism, Daoist Dao-de Individualism and Buddhist Internal Relationalism (Callicott 2014: 386).

Notwithstanding these attempts at establishing a space for both, environmental ethics and cross-cultural ruminations, ongoing debates in environmental ethics and aesthetics, especially in the German-speaking countries, continue to, in general, circumvent these positions. Partly, this has to do with one particular issue at the core of environmental ethics: the value of non-human nature. As is well known, axiology or *Wertlehre,* which was established as a philosophical discipline at the end of the nineteenth century in the post-Kantian German philosophical landscape, is based upon tacit, non-universal assumptions regarding the relation of a subject to an object. These tacit assumptions continue to remain uninterrogated, epistemic anthropocentrism continues to be the standard position in the main environmental moral theories in the German-speaking countries. Attempts at critiquing this anthropocentrism are met with reservations, sometimes even hostility. Arguments which underscore the inherent, or even objective, value of nature, as well as those that attribute rights to non-human living beings or ecosystems are, in general, not regarded as serious philosophical alternatives.

Moreover, at least two factors seem to hamper a sustained discussion on environmental concerns: typical, entrenched distinctions between a moral, aesthetic, scientific and religious attitude to nature, which is again rooted in modern Euroamerican philosophy, and a reductive view of aesthetic value, which bases it on positive human emotions. Arguably, such distinctions may indeed serve theoretical purposes. They are, however, counterintuitive and contradict our common experience of nature in our everyday lives. Finally, one could ask whether environmental ethics is necessarily bound up with a teleological model of action which imposes norms and obligations.

If publication output and conference attendance are an indication, scholars located in Asia and Africa and/or those with links to these regions seem to show an increased interest in environmental ethics and aesthetics. Their participation has yet to be reflected in German publications on environmental ethics. In addition, German literature on intercultural philosophy tends to focus on issues of ontology, history of philosophy, aesthetics, logics, political philosophy, and general ethics. One upshot of this trend is that the discussion of environmental issues in Europe is restricted either to the continent's own past (e.g. Paracelsus, Spinoza, the Romantic philosophy of nature, etc.) or to an analysis of models developed in Europe and North America. As a consequence, environmental ethics is characterized by an asymmetry between debates on moral-philosophical theories in the Euroamerican academy and information related to other cultures, hereby continuing the trajectory of a decidedly Eurocentric anthropocentrism, even if inadvertently.

As feminist theorists have been arguing for some time now, this anthropocentrism is closely related to androcentrism. Early feminist literature on environmental philosophy tended to underscore the interconnections between women, animals and nature. These early reactions to the nature/nurture, nature/culture divide were soon confronted by the critique that the »universal woman« who informed these writings was an able-bodied, upper-class, white, academic female speaking for (or claiming to represent) the rest of womankind.

By decentring her perspective as the sole female perspective of scholarly worth, the ensuing methodological correction opened up the field to a host of pertinent issues. The list of possible oppressive factors to women were not simply restricted to those which were experienced by these privileged women in their affluent (welfare) states. It also brought into scholarly focus new ways of conceptualising the relationship between (female) human beings and their environment. One upshot of this recalibration has also been research on new materialisms. Some work in this area draws attention to how a corporeal self itself is constituted by relations to other non-human selves. Not only are we not singularly male or female, we are a mass of swarming microorganisms within the coordinates of a single body. Some postcolonial STS scholars go further and argue, if beliefs in a »universal man« or a »universal woman« have been effectively debunked, why not open oneself up to the possibility that analogously there could perhaps be more than one understanding of matter which constitutes the body?

The scale of our environmental crises is undeniably global. Environmental problems do not seem to respect conventional boundaries drawn by nation-states or communities. In fact, there are indications that some such problems (plastic pollution in the seas or even climate change) can be dealt with effectively only through international cooperation. In addition, this cooperation seems to be necessary in solving thorny issues of biopiracy, setting up national parks and wildlife reserves across state borders, investments in the development of an environmental-friendly technology, and the protection of Indigenous traditional knowledge that would limit the profit of multinational corporations. A mutual dialogue is needed especially when humans' needs are pitted against the protection of threatened environments or non-human species, like in less-affluent countries, where economic growth has to cope up with demographic issues. In particular, a new thinking is necessary in order to share the responsibility for climate change and perhaps even assign more tasks to more-affluent countries. In addition, the sheer scale of our environmental crises suggests that dominant patterns of living and consumption may urgently need radical rethinking in the decades to come. On all these fronts, an international participation of philosophers seems to be warranted for a more nuanced philosophical reflection about global solidarity and global justice.

As papers in this volume indicate, this work is already underway. They illustrate that a much-needed, novel, more daring way of reconceptualising our relationship with nature is indeed possible. These papers also highlight how positions from other traditions may provide theoretical frames that would help to overcome the subjective character of value judgments. These, in turn, could be fruitful in establishing norms on a global scale that could possibly enable a sustained protection of endangered environments. Let us turn to these individual papers now.

1 Creative Explorations

In rethinking issues pertinent to environmental ethics, one could argue that our current environmental crisis is so grave, that all resources available to humanity should be harnessed in offering a fine-grained analysis of the crisis as well as developing viable responses to the same. Scholarly engagements will be broad in scope only by drawing upon these resources. Additionally, such engagements will be able to appeal to those located outside the bounds of the academy too. One way of doing so would be through an integrative analysis which focuses on hitherto occluded aspects within an approach or understudied similarities between approaches.

Hava Tirosh-Samuelson turns to this task in her ›Ethics of Care and Responsibility: Bridging Secular and Religious Cultures‹. Arguing against the claim that the Judeo-Christian tradition licenses an unbridled domination of nature, Tirosh-Samuelson offers an ecological reading of the biblical message. By revisiting the Judaic ethics of responsibility, she sketches the inter-relational aspect of all life. Human beings have duties toward all creatures on account of being a human created by God. The principle underlying this ethical stance places human beings in the unique position of caring for God's creatures and being responsible to God for this care. As we see, this position seems to offer an account of environmental justice which is grounded in duties. But this is not all. Tirosh-Samuelson directs our attention to the link between the Judaic ethics of responsibility, Zionism, an experience of the outdoors in Israel and the development of a secular understanding of Jewish environmentalism in Israel and North America today.

Despite having bridged the gap between the Judaic narrative and more secular concerns, Tirosh-Samuelson notes one lacuna: Early Jewish environmentalism did not fully engage with philosophical accounts of care ethics, although these were partly developed by secular Jewish feminists and philosophers, and vice versa. Despite the common focus on relationality, vulnerability, responsibility and care, both these factions worked in relative isolation from each other. The succeeding generation of Jewish-born eco-feminists were indeed able to bridge this gap. In their interpretation, women tend to be portrayed as paradigmatic care-takers of the Earth. In addition, these eco-feminists understand Earth-care as a spiritual task. A Judaic grounding of ethical values, and their inculcation, would be one viable way to adequately tackle the global dimension of the environmental crisis, she argues.

In her ›Christliche Umweltspiritualität als Antwort auf die Umweltkrise‹ [Christian Environmental Spirituality as an Answer to the Environmental Crisis], Ingeborg G. Gabriel complements Tirosh-Samuelson's discussion from the Christian perspective. Her article explores and problematises certain interpretations of biblical precepts, which in a standard interpretation have been made culpable for the derailment of modern economy, technology, and ethics. One salient precept in this regard is God's commandment to the first human beings to »be fruitful and multiply and fill the earth and subdue it and have dominion […] over every living thing that moves on the earth« (Gen. 1:28). Placing this command in its historical perspective, Gabriel criticises the simplistic notion of a linear development and continuity between the epoch of the Bible and our times. This notion, in her view, is simplistic insofar as it ignores the process of our gradual alienation from nature as well as the exegetical distortions of the aforementioned commandment. Like Tirosh-Samuelson, Gabriel too underscores that the Judeo-Christian tradition perceives human beings as preservers (or carers, as Tirosh-Samuelson puts it) of a divine creation which is in perpetual change. From their different perspectives, both authors home in on virtues highlighted in the Judeo-Christian tradition, like humility, moderation, deliberate abstention, simplicity, self-control, as also thankfulness. These virtues form an important link to debates on green virtues in environmental ethics.

However, virtues remain abstract without developing skills and being materialised into practices. In his ›Rituals as Environmental Skills: Inhabiting Place, Fabricating Meaning, Enhancing Morality‹, Sigurd Bergmann makes a case for using the power of rituals to initiate hitherto unexplored ways of relating to the environment. Bergmann attempts to show that the encounter between ritual studies and aesth/etics has the potential to transform, or rather initiate a transition toward a more environmental-friendly »ecocene«. As place-based social acts of normative significance, rituals can transform the broader environment in which they are carried out. Their normative power derives from their ability to thematize ultimate concerns, which, in turn, are expressed through ultimate sacred postulates. These postulates have the power to deflate the human *hybris*. Inasmuch as they link up the local with the cosmic, the rituals symbolising them work best with a synthesis of rational, ethical and aesthetic dimensions. However, Bergmann argues that it would be far-fetched to claim that all rituals per se promote so-called green virtues.

Rituals which revolve around modern economic and technological power tend to impact the environment negatively. In fact, these rituals seek to maintain the status quo, and with it, human domination of nature. Only with an embodied perception, only with an »aest/ethics«, can we obtain a deeper awareness of what is morally demanding. Only through this awareness can we deepen common understandings, provoke new ones, and critically transform the ways in which we as communities relate to the natural environment, even in the ritualised spaces of our daily lives.

How would cross-cultural explorations in environmental ethics help to enable the radical rethinking alluded to above? Arguably, a conceptual analysis which deconstructs basic concepts like ›ethics‹,

›nature‹, or ›environment‹ would be one way to begin. Their standard rendition in academic philosophy tacitly takes over assumptions of European philosophy, hereby not adequately taking into account the Greek, Latin or French origin of such concepts. Additionally, this work can be complemented through the introduction of concepts from other cultures, which can enrich, put into perspective or even replace the conventional, academic understanding of the so-called ›natural environment‹. Such work can illustrate alternative ways of reconceptualising the relationship between a human being and the environment. Roman Paşca's ›The Self-With-Others and Environmental Ethics‹ does precisely that. Paşca engages with the relation of *hito* (human being) and *shizen* (realm of energy) as was developed by the Japanese philosopher Andō Shōeki 安藤昌益 (1703–1762). In this interpretation, each human being is an ontological combination of man and woman. One *hito* is in this manner also related to another; being interrelated, we are all mutually interdependent. Shōeki's *hito* is ontologically linked with the universe too.

In a pristine, primordial state, all manifestations of energy and life are uncorrupted. This is the *shizen no yo*, which is distinct from the *shihōsei*, the world made by human beings. Through our male-female *hito*, we partake in the *shizen*. More precisely, we, as human beings, are related to the *shizen* through the *gosei* (互性, »mutual natures«). Narrowly understood as nature, *shizen* in this interpretation is to be understood as a kind of spontaneous being and becoming. It is the way in which the world presents itself to us. Seen from the perspective of *shizen*, human beings are ahistorical and timeless; we are all part of that one, eternal energy. Within its realm, it would be meaningless to postulate hierarchies. Paşca then goes on to develop Shōeki's novel way of harnessing human potentialities. With a creative understanding of *chokkō* (»straight cultivation«), Shōeki intends to channel human activities and align them with the creative energy of *shizen*. The idea here seems to be that through such a ›straight cultivation‹, every mundane aspect of our lives would be infused by the primordial energy, which connects all beings. Living, or rather in striving to live, such a life, we would realise that there is no ontological, or even epistemological, divide between us and our environment. We are all one and the same. Shōeki's interpretation of eternal energy and life would be one way of reconceptualising the relationship between a human being and the environment. Put in our own terms, it offers a non-anthropocentric, non-androcentric perspective on this relationship.

Ted Toadvine's ›Climate Apocalypticism and the Temporal Sublime‹ complements some aspects of this discussion, albeit from a different perspective. Toadvine invites us to radically rethink our conceptual frameworks. Especially, in societies dominated by Judeo-Christian thought, Toadvine sees a tendency to adopt the apocalyptic structure in thinking about climate change. However, insofar as this structure is wedded to the narrative of a future catastrophe, it fails to exploit the creative tension engendered by apocalyptic thinking. In the linear understanding of time this thinking adopts, the crisis leads up to Judgement Day. This day is projected as that specific moment in an unforeseeable future at which the larger ramifications of our actions will be made clear to us. However, this projection seems to rob us of our potential for action here and now. Its rhetoric proves to be ineffectual and deterministic; it is unable to galvanise us to undertake appropriate measures to combat climate change now.

Toadvine is determined to refuse the rhetorical power of Judgement Day. Reading Jean-Luc Nancy's call to think the »ever-present« along with Potawatomi scholar Kyle Powys Whyte's proposal of »time-spiraling«, he seeks to develop a more holistic perspective of the present. Should we be able to locate the crisis in our present, he argues, this would appeal to our human yearning for »transcendental significance«, for the »deep temporal sublime«. By bringing the future in the present, as it were, we would be able to break the linearity of time, which is commonly understood as a chain of ever-substitutable moments. Foregrounding the temporal sublime, we would then make ourselves aware of the unique and non-substitutable moments which constitute our lives. We would then realise how our own temporal materiality binds us with an inter-species past, and future.

This is where Whyte's ruminations of temporal justice become relevant. Indigenous experiences of ›spiraling time‹ set the protagonists in an ever-present dialogue with ancestors and descendants. Such experiences are important reminders of the fact that our world of colonial violence and climate change are not outside our own time. In our singular present, they bind our pasts with our futures.

2 Cross-Cultural Explorations

Cross-cultural explorations in environmental ethics can bring to light broader issues of significance for specific societies too. As a common charge will have it, the environmental crisis is directly related to the subject-object dichotomy and to the forgetfulness of human beings to their own »nature«. As a result, there is a tendency for humans to act as if they were separated from their living environment, including other people, and behave as if they could act intentionally without being affected bodily. One alternative would be to retrieve an awareness of one's own finitude and a relational thinking in a universally interconnected world. A relational self acknowledges its interdependence with the physical and social environment, in which humans, animals, plants, physical landscape, and weather forces are interwoven and sometimes animated. The awareness of an extended kinship —with living beings or even with cosmic forces and elements—and the replacement of anthropocentrism with a thinking in terms of sym-biosis and interrelatedness are two answers to the ecological crisis that can be found not only in Asian philosophical traditions, but also in other cultures. In the present volume, Silvia Donzelli's and Mădălina Diaconu's papers on African and Romanian traditions respectively bring to light unintentional overlappings that make visible more general problems of cross-cultural explorations into environmental philosophy. Two of these problems regard the ambivalence of ethnophilosophy and the discrepancy between theory and praxis. The former sets up an opposition between idealised folk traditions and colonial hegemonic cultures and/or Euroamerican modernization in order to support a politics of identity and liberation; the latter deals with the clash between traditional environment-friendly worldviews, as emphasised by ethnophilosphy, and real economic needs and practices.

Silvia Donzelli's ›Umweltethische Ansätze in der afrikanischen Philosophie‹ [Approaches to Environmental Ethics in African Philosophy] begins by noting the link between African environmental philosophy and *Négritude*. The latter, as we know, began as an active form of pan-African resistance to colonial domination in the urban spaces of colonial Paris in the early 1930s. The ensuing debate on environmental ethics in Africa picked up some themes that had already been broached by the thinkers associated with *Négritude*. One such theme was a pristine and harmonious relationship between human beings and nature in precolonial Africa. This was, in Donzelli's analysis, related to the development of a so-called ethnophilosophy—considered to be an autochthonous way of philosophising of the African peoples. In both cases, the precolonial past was perceived to be free of the evils of the present. In the initial debates on environmental ethics, African scholars tended to attribute an instrumental view of human rationality and the domination of nature to the onset of colonialist practices on the continent.

This way of premediating the relationship of a present-day African to nature through colonialism is shifting gradually. To underscore her point, Donzelli briefly sketches the work of three contemporary environmental ethicists: the Ethiopian philosopher Workineh Kelbessa, the Nigerian philosopher Segun Ogungbemi and the South African-American philosopher Thaddeus Metz.

In different ways, all three authors seem to distance themselves from the erstwhile romanticised view of precolonial Africans' relationship to nature. Their distinct approaches operate with a relatively normative view of nature. Kelbessa does not seek to develop one, overarching principle driving environmental ethics. He argues that we should respect the local groups' (like the Oromo) attempts to deal with modern technology. Ogungbemi does indeed attempt to work out an overarching principle drawn from traditional African ethics, which could however be applicable to today's environmental concerns. This principle of »nature-relatedness« is said to be independent of any particular ethico-religious grounding. The last in the trio, Metz, attempts to chart a territory between holism and individualism. Leaning on *ubuntu,* Metz attempts to show how its descriptive-moral quality can be used to ground a larger argument about an attribution of intrinsic moral status to entities. Donzelli concludes the paper by briefly sketching how these positions would relate to some aspects of the environmental devastation caused in the Niger delta by the production of crude oil.

Mădălina Diaconu's paper ›Von prämoderner Naturliebe zum (trans)nationalen Umweltaktivismus: Der Fall Rumänien‹ [From a Premodern Love of Nature to (Trans-)national Environmental Activism: Romania as a Case] continues Donzelli's discussion about the ambivalence of what is called ethnophilosophy and the discrepancy between theory and praxis. Diaconu's paper sheds light on how a cross-cultural environmental philosophy may pose a dilemma for non-Euroamerican philosophers associated with the traditions they engage with as scholars. In some cases, they have to tread a fine line between praising their own cultural traditions and working to increase an awareness for environmental issues using debates established in Euroamerican philosophy. While they strive to avoid nationalism in their praise of their traditions, they have to, equally, circumvent the onto-theological background of the Euroamerican tradition.

Diaconu's paper argues for a greater awareness of the situated character of philosophising and of the difference between an inner and an outer perspective in cross-cultural debates. The latter is meant to capture the view of those who, not standing in a concrete relation to the non-Euroamerican tradition, still use it to question the hegemony of the Euroamerican tradition. The former captures the perspective of those organically related to the non-Euroamerican tradition studied. Given the current asymmetry in cross-cultural philosophical explorations, these scholars are usually expected to present non-Euroamerican traditions in a favourable light, which however weakens from the start any possible efficient self-critique of their country of origin. Moreover, non-Euroamerican scholars have to work with polyvalent messages, one for the in-group and one for the out-group. Diaconu uses the example of Romania to make this point.

A holistic and basically premodern environmental-friendly worldview is essential for the self-understanding of Romanian culture. A major contribution in this respect is owed to Ovidiu Papadima (1909–1996) and Mircea Vulcănescu (1904–1952) before WWII, followed by Constantin Noica (1909–1987) during national communism. For example, Vulcănescu and Noica emphasised the polysemy of the Romanian word for ›nature‹ and reconstructed the popular (for Papadima also Christian) representation of an interconnected living universe in which the visible communicates with the invisible, and the real is surrounded by the possible on the basis of language and folk traditions. However, their ontology influenced by phenomenology remained without consequences for the Romanian economy on its way to modernization, and the blatant divergence between theory and practice is still a blind spot in Romania. Despite the intensive reception of anglophone philosophy and the explosive development of applied ethics after the fall of communism, environmental philosophy and environmental ethics continue to be neglected, even in feminist philosophy. The intensified degradation of the environment in the context of the confused transition to a new political order as well as the recent protests against the selling of Roșia Montană to a Canadian mining corporation catalysed the maturation of the civil society, yet in the flagrant absence of corresponding philosophical debates. Against this background, the reception of Euroamerican environmental ethics may be assigned a positive role in raising awareness for environmental issues before the emergence of context-specific theories. Adrian Miroiu, who positions himself outside the divide between the so-called environmental ethics and Indigenous or other »exotic« worldviews, as he calls them, has already made a beginning in this regard. Arguing against an undifferentiated rejection of Judeo-Christian thinking, Miroiu emphasises the potential of Eastern Christianity for grounding a virtue ethics that avoids prescriptions imposed by the state (that would be less acceptable in a society that suffered for a long time from authoritarian regimes) and postulates the intrinsic value of nature.

Stefan Knauß' paper ›Planetarische Integrität: Was Umweltethik und interkulturelle Philosophie voneinander lernen können‹ [Planetary Integrity: What Environmental Ethics and Intercultural Philosophy Can Learn from Each Other?] is a good example of the so-called outer perspective identified by Diaconu. Knauß attempts to take forward the project of German-speaking intercultural philosophy against the backdrop of international legal developments, in which rights have been attributed to nature. If, he asks, courts have been open to, and accepted, particularistic groundings through which natural entities have been attributed rights, why cannot philosophy dock onto these developments in its role of being part of »western science?«

In his ambitious sketch of a ›planetary integrity‹ *(planetarische Integrität)*, Knauß uses phenomenological sources to ground the experience of unfamiliarity. In our encounters with members of other cultures, we, according to this argument, may experience similarity, but also moments of stark difference. The challenge is to strike a right balance between the two. To maintain this balance, it seems, a normative position is necessary, which can seriously take into account the embodiment, the exteriority (»Exteriorität«) of the other person. Knauß locates this position in the philosophical debate on integrity. Through our respect for other embodied persons and their specific cultures, we can arrive at the notion of a planetary integrity. This notion would allow us to conceive of Earth as a shared space, a common *Lebensraum* for all (living) beings. In this manner, Knauß hopes that German intercultural philosophy will serve as a vehicle to pave the way for a more inclusivist understanding of nature in German-speaking debates about environmental ethics.

3 Socio-political Explorations

Tatiana Mishatkina's paper ›Environmental Ethics and Environmental Security: Specifics of the East-European Region‹, provides an overview of developments in the field of environmental philosophy in the Post-Soviet Space, which includes Russia, Ukraine, Belarus, Lithuania, and the Republic of Moldova. Mishatkina enumerates the activities of key institutions involved in environmental research and education in these states. This much-needed overview sets in with an analysis of political, ideological, and psycho-moral factors that hindered the reception of Euroamerican environmental ethics during the Soviet era. These factors range from a deep suspicion toward theories emanating from the capitalist world to the trauma left behind by the explosion at the atomic plant in Chernobyl in 1986.

Mishatkina argues that a »post-humanity«, perspective which involves care of all species, is necessary for the whole world, including states located in Post-Soviet Space. But how is this need perceived within these states themselves? Using the grounding of environmental centres at universities and other state research institutions, international conferences, publications, scholarly journals, as well as projects carried out by NGOs as points of orientation, Mishatkina reconstructs the history of environmental ethics in this space over the past twenty years. She also familiarises us with the pioneering work of R. G. Apresyan, whose take on general and practical principles of the environmental ethics mirrors those circulating in international literature, be these respect for all forms of life, biodiversity, sustainability, environmental justice, the precautionary principle, etc.

Mishatkina's paper does more than summarise debates on environmental ethics in post-Soviet regions. She uses the overview to ground her own position on environmental ethics. This position seeks to charter a middle course between theoretical principles on one hand and practical guidelines on the other. Her endorsement of »subject-subject relationships« parallels Toadvine's argument for grounding the future in the present. For the sake of future generations, we have to move away from a human-centred perspective, she argues. Ecosystems have an inherent value. Mishatkina holds that her conceptual model is appropriate even for post-Soviet states like Belarus, which for historical reasons did not have an extensive debate on environmental ethics in the past. She concludes her paper by focusing on educational projects, which work toward establishing an environmental-friendly view in Belarus.

Cheng Xiangzhan's paper ›Ecological Civilization and Ecological Aesthetics in China: An Overview‹ endeavours to enunciate the concept of an ecological civilization. Drawing on Iring Fetscher's work, Chinese scholars began to use the term in the 1980s to warn about an over-optimistic view of economic development from the perspective of ecological consciousness. In 2007, the Chinese government proclaimed ecological civilization, in Chinese ›*shengtai wenming*‹ (生态文明), to be a development goal. Cheng Xiangzhan notes how Chinese political documents explicate this concept and embeds it in international literature on environmental ethics. Keen to emphasise the originality of contemporary Chinese approaches in the field, he sketches the history of ecological aesthetics in the United States and its late reception by Chinese scholars.

Referencing his own work, as well as that of Zeng Fanren, Cheng Xiangzhan works out two salient perspectives on ecoaesthetics in the Chinese context. Zeng Fanren, Cheng Xiangzhan notes, initially reconstructed the »postmodern« context in which the debate on ecological aesthetics emerged. At this stage, he outlined a so-called »aesthetics of ecological existence« based upon the work of Martin Heidegger. However, since 2005 he, like Cheng Xiangzhan himself, explicitly seeks to relate ecological aesthetics to the concept of ecological civilization. Unlike environmental moral philosophers who reject humanism due to its anthropocentric »arrogance« (David W. Ehrenfeld), Chinese authors working on ecological aesthetics do not give up the claim of humanism. Rather, they aim to construct a »civilization in harmony with nature«, in which humans, conceived as embodied subjects, respect nature, act according to it and protect it. As Cheng Xiangzhan notes, this understanding of eco-aesthetics, which he distinguishes from environmental aesthetics (a concept coined by Arnold Berleant in the US), is both scientific and philosophical. In addition, it draws upon philosophical sources located within China, Europe and North America which highlight the consonance between humans and nature.

Issues of conservation and preservation, which are alluded to by both Mishatkina and Cheng Xiangzhen, are at the centre of Dean Brink's paper ›Rewilding and Neoliberal Territorialities after the Anthropocene: Cybernetic Modelling of the Oriental Stork as Critique‹. Critiquing conservation attempts of the oriental white stork *(Ciconia boyciana)* in the Japanese Hyogo Prefecture from a Deleuze-Guattarian perspective, Brink endeavors to illustrate the merits of foregrounding »posthuman cybernetic dependencies«. According to this analysis, extant Japanese attempts at rewilding the oriental white stork are bound to fail given that these attempts are part of the complex neoliberal enterprise of exploiting nature and hereby increasing goods for human consumption. Brink observes that this neoliberal dimension is papered over by the conventional understanding that they, members of Japanese society, have a long history of preserving an intact relationship with nature; past malformations of the beauty of the Japanese landscape are generally attributed to foreign influences, thus paralleling Donzelli's remarks about ethnophilosophy in the African context. This naïve view further occludes the neoliberal impact on nature.

In this densely-argued paper, Brink rests his argument for a radical rethink on a posthuman perspective, which alone can take a clear stand on nonhuman-human-nonhuman relations. The »species interaction« afforded by this perspective would be one viable way out into a post-Anthropocene world. Brink envisions cybernetics—meaning, a multi-channelled recognition of inter-species needs as well as human affects—as mitigating the effects we, human beings, have on other species. Brink views Deleuze and Guattari's understanding of cybernetics as having the potential to decentre human beings from the pyramid of life and enable a deliberate taking-back of human importance such that non-human entities can relate to each other too.

One interesting point of contrast to Tatiana Mishatkina's and Cheng Xiangzhen's papers has to be noted here. Brink does not endorse an uncomplicated going back to philosophical roots, if you will. He attributes the overemphasis on human intentions and needs to idealism borne in the period of imperialist expansion in Japan, a discourse supported in various ways by Watsuji Tetsuro as well as the idealist tendencies of the influential Kyoto School. Such philosophers, writes Brink, leaned too heavily on theories of human intentionality developed by the likes of Kant, Hegel and Bergson. For Brink, the long-term results of adopting this perspective, which does not contain a vison of a non-human centred world, are upon us today. Philosophical positions from the Kyoto School, he implies, may not be sufficent to adequately address the rice-destroying activities of the stork.

A relational self, relational virtues, relationality among species, interdependency, the intrinsic value of nature, Earth-care, ethnophilosophy, time in-between, a posthuman world etc. are some of the threads running through this volume dedicated to a cross-cultural exploration of environmental ethics and aesthetics. We hope that it will be able to initiate a more nuanced perspective on global solidarity and justice as these relate to our relationship with the environment.

—Monika Kirloskar-Steinbach and Mădălina Diaconu

References

J. Baird Callicott, »Non-Anthropocentric Value Theory and Environmental Ethics«, *American Philosophical Quarterly*, Vol. 21, No. 4, 1984, pp. 299–309.

J. Baird Callicott and Roger T. Ames, *Nature in Asian Traditions of Thought: Essays in Environmental Philosophy*, New York, 1989.

J. Baird Callicott, »Afterword. Recontextualizing the Self in Comparative Environmental Philosophy«, in *Environmental Philosophy in Asian Traditions of Thought*, ed. J. Baird Callicott and James McRae, New York, 2014, pp. 377–390.

ed. J. Baird Callicott and James McRae, *Environmental Philosophy in Asian Traditions of Thought*, New York, 2014.

David W. Ehrenfeld, *The Arrogance of Humanism*, Oxford, 1981.

Eugene Hargrove, *Foundations of Environmental Ethics*, Englewood Cliffs, 1989.

Gerald James Larson, »›Conceptual Resources‹ in South Asia for ›Environmental Ethics‹ or The Fly Is Still Alive and Well in the Bottle«, *Philosophy East and West*, 1987, Vol. 37, pp. 150–157.

Creative Explorations

Ethics of Care and Responsibility: Bridging Secular and Religious Cultures

1 Religion and the Eco-Crisis of the Anthropocene

We live in the midst of a massive anthropogenic ecological crisis whose manifestations include global warming, shifting weather patterns, extreme weather events, retreat of glaciers, rising sea levels, mega-droughts and desertification, threats to available water, food and shelter, mass extinction of species, loss of fisheries and forests, acidification of oceans, pollution of air, water, and soil, and shifts in the range and prevalence of diseases (Henson 2014). The anthropogenic eco-crisis marks the dawn of a new geological age, the Anthropocene, a new period in geological time that marks human transformation of environmental systems (Steffen et al 2011; Ruddiman 2013; Ruddiman et al 2015; Williston 2015; Angus 2016; Bonnheuil and Fressoz 2017; Ellis 2018). In the Anthropocene humankind has become a geophysical force as the boundary between »nature« and »humanity« has collapsed when »people and nature interact reciprocally and form complex feedback loops« (Liu 2007: 1513). Currently, all forms of life on planet Earth are impacted by human activities permanently and irreversibly. The temporal scope, causes, and implications of the Anthropocene are still disputed as the Anthropocene »has become a contentious term and a lightning rod for political and philosophical arguments about what needs to be done, the future of humanity, the potential of technology and the prospects for civilization« (Dalby 2016: 34).

How should humanity respond to the new reality it has created? What should guide the deliberations about the technological, economic, social, political, and moral choice it will face in the Anthropocene? Can modern technoscience provide the guide to the perplexed if technoscience is the primary cause of the crisis we now face? These complicated questions are hotly contested, but as Mike Hulme has already noted in regard to the debate on climate change, »what is contested are not observable facts and physical realities but the *interpretations* of those facts, namely, the meaning we ascribe to them, and the cultural, social, political and ethical practices that flow from them« (Hulme 2009: 147). The global eco-crisis poses social, cultural, and ethical questions that require humanity to engage its deep moral, religious and spiritual resources as it charts its responses. Hulme has already noted that »climate change is increasingly discussed using language borrowed from religion, theology and morality« (Hulme 2009: 173) and indeed the relevance of religion to human coping with the Anthropocene has been increasingly acknowledged (Kearns 2011; Clingerman et al 2014; Deane-Drummond et al 2018).

Religion is the repository of sacred narratives that frame our worldviews, morality, attitudes, and practices. Religion matters most to the environmental crisis because the overwhelming majority of people in the world conceptualise reality in religious categories (Grim and Tucker 2014: 28). Human beings understand themselves, their societies, and their daily life through sacred narratives and symbolic rituals that point beyond themselves to an ultimate reality. In the context of religious worldviews human beings organise their life and find meaning, purpose, and hope as they face an unknown future. Religion provides the moral lens through which humans evaluate every aspect of life and decide what is good and bad, what is permitted and forbidden. Religion also expresses human existential and emotional needs and frames—that which we care most about, namely, our ultimate concern. Because religion expresses human orientation toward ultimate reality, humans are willing to kill and die for it so that religion mobilises people to action more than any other factor. If we are to mitigate the environmental degradation by changing human conduct, we cannot leave religion out of environmental discourse. Indeed, the depth and scope of our eco-crisis demands that we reframe it in religious categories, because no other dimension of human life (e.g., science, law, or philosophy) is more compelling than religion. Religion, however, is not a set of fixed beliefs or dogmas but a comprehensive way of life that encompasses human attitudes toward time, space, place, embodiment, sex and gender, family, community, and ultimately, life and death. Judaism, a small but foundational monotheistic religious tradition, is a case in point.

Judaism articulates an environmental ethics of care and responsibility that addresses our contemporary eco-crisis, but it does so by cutting across the boundary between »religious« and »secular« cultures. That Judaism is environmentally relevant is not uncontroversial because the Judeo-Christian tradition has been blamed for the current ecological crisis (White 1967). According to Lynn White Jr., the Bible (Gen. 1:28) gave humanity the mandate »to have dominion« over the Earth, using and even exploiting its natural resources for human benefits. Furthermore, White has charged that Christian dualism of body and spirit has facilitated the disenchantment of nature because it made possible the rise of modern science and technology which brought about the ecological crisis. »Disenchantment« or »demagification« *(Entzauberung)* is the term coined by Max Weber (1917) to denote the fact that in the modern world nature was no longer regarded as the abode of divine presence but is rather viewed as inert matter that can be manipulated and mastered for the benefit of humanity. In the modern secular worldview God became increasingly irrelevant to the understanding of life on Earth and was replaced by the human who was now seen as »the creator and arbiter of values [who] is free to interpret and manipulate nature as it pleases« (McGrath 2002: 54). The replacement of God by humanity was the core of the modern secularization and its Enlightenment faith in the perpetual »progress« of humanity by means of science and technology. Along with other forces of modernity—capitalism, colonialism, and imperialism—the Enlightenment's ideology of progress yielded massive devastation of nature as well as the subjugation of human populations who were viewed as mere »resources« for empowerment and enrichment of Western »enlightened« nation-states (Anker 2001; Hedrick 1988; Tucker 2000).

Within three centuries of rapid technological development, geopolitical expansion, and exploitation of nature, modernity has resulted in massive environmental degradation and the collapse of Earth systems we now associate with the Anthropocene. Life on Earth has become so precarious not because the Judeo-Christian tradition gave license to humans to exploit nature, as Lynn White has argued, but rather because the ecological message of the Bible that humanity is responsible for the natural world and must take care of the world has been occluded or ignored. The responses of theologians, scholars of religions, and activists to White's accusations have given rise to the discourse of religion and ecology which consists of ecological hermeneutics, eco-theology, comparative analysis of religious traditions, and religious environmental activism (Hargrove 1986; Taylor 2005; Gottlieb 2006; Jenkins, Tucker, and Grim 2017; Hart 2017). All world religions, each in its own unique way, harbor profound ecological wisdom that guides humanity to care and protect the natural world. In the Abrahamic traditions ecological wisdom is rooted in the doctrine of creation according to which God is the sole creator and sustainer of nature and God has commanded humanity to care of the created world that ultimately belongs to God. Especially in America, »religion is the most likely way that Americans can move themselves to care for the Creation« (Oelschlager 1994: 75). The Judeo-Christian tradition then does not mandate human domination and control of nature but calls humanity to care for creation so as to protect the divinely created world (e.g., Habel 2000; Berry 2006; Horrell 2010).

2 Jewish Environmentalism: Religious and Secular

Jewish environmentalism emerged in the early 1970s in direct response to Lynn White's claims that the Bible is the cause of the environmental crisis (Helfand 1971; 1986; Lamm 1972). The initial response was indeed apologetic, arguing that White selectively focused on the biblical command to »subdue the Earth and have dominion over it« (Gen. 1:28), which seems to justify human mastery and control of nature, while ignoring the command to the human »to till and protect« the Earth (Gen. 2:15), which sets a clear task for humanity. White's short essay focused on Genesis to the exclusion of the rest of the Bible and it had nothing to say about rabbinic Judaism, the normative interpretation of the Bible, or post-rabbinic developments of Judaism. Contrary to White, Jewish theologians and exegetes noted the richness and the complexity of the biblical portrayal of nature and the role that nature plays in covenantal relationship between God and Israel (Kay 1988 [2001]; Artson 1991 [2001]). Jewish environmentalism is engaged in the recovery, retrieval, and reconstruction of the Jewish tradition in light of the values and attitudes of contemporary environmentalism: interconnectedness of all beings, sustainability, concerns for future generations, and compassion for nonhuman animals (Tirosh-Samuelson 2002). A dominant value of the movement is the commitment to *Tikkun Olam*, literally »the repair of the world«, a rabbinic idea that evolved over the centuries and has become the catchphrase for Jewish social activism (Troster 2008). In rabbinic Judaism the phrase expressed the eschatological hope for the coming of the Kingdom of God. In medieval Kabbalah, the concept was reinterpreted to refer to the transformation of all levels of reality by means of theurgy performed with the proper intention. In modern Judaism, especially in Reform or Liberal/Progressive variants, *Tikkun Olam* pertains to the range of human activities that bring about social justice so as to improve the human condition. The Jewish environmental movement has adopted the concept of *Tikkun Olam* interpreting it to mean a fusion of social justice and eco-justice through a wide range of environmental activism.

The Judaic ethics of responsibility is inherently religious: humans are responsible *for* the well-being of the created world and responsible *to* God for the treatment of the divinely revealed created world. The care of God's created world flows from human responsibility to God which is framed not in terms of the intrinsic *rights* of nature but as human *duties* toward the natural world which God had created (E. Schwartz 1997 [2001]). Scriptures spell out the ways in which humans should treat the land, water, plants, animals, and other human beings in order to protect God's created world (Richter 2010). In the Scriptural worldview, God the Creator is the true owner of the Earth and humans are but temporary tenants. Therefore, Scripture commands Israel to give the first portion of the land's yield (Hebrew: *bikkurim*) to its rightful owner in order to ensure the land's continuing fertility and farmer's sustenance and prosperity. Scriptural land based-commandments (e.g., Lev. 19:23) pay attention to trees and command that during the first three years of growth the fruits of newly planted trees and vineyards are not to be eaten because they are considered to be God's property. Scripture also recognises the goodness of biodiversity, protects diversification and prohibits mixing different kinds of species of plants, fruit trees, fish, birds and land animals (Hüttermann 1999). The biblical concerns for animal welfare are evident in the stringent limits on human consumption of animals by differentiating between »clean« and »unclean« animals, the concern for the perpetuation of life of non-human animals, and the prohibition on causing unnecessary suffering to animals (Tirosh-Samuelson 2017).

The most distinctive feature of Jewish environmental ethics is the causal connection between the moral quality of human life and the vitality of God's Creation. Conversely, the moral corruption of society evident in unjust treatment of the marginal elements of society is closely linked to the corruption of nature. In both cases, injustice arises from human greed and the failure to protect the original order of Creation. The just treatment of the marginalised in society—the poor, the hungry, widows, orphans—must follow Scripture's legislation (Deut. 15:1–7). Thus, parts of the land produce are to be given to those who do not own land. When people observe the commandments, the soil itself becomes holy and the person who obeys the commandment ensures the religious-moral purity necessary for residence in God's land (Deut. 4:40). Failure to treat other members of society justly and to protect the sanctity of their lives is integrally tied to acts extended toward the land. Put differently, there is no tension between the principle of justice and the practice of care, a tension that concerned contemporary secular moral philosophy and its feminist critics.

The connection between land management, divine worship, and social justice is most evident in the laws regulating the Sabbath and the sabbatical year (Hebrew: *shemitah*). On Sabbath, creative work is prohibited in order to enable humans to devote time to reflection and recognition of human subservience to greater power. Rest is imposed not only on humans but also on the domestic animals in their service; environmental legislation impacted animals even though they were not directly commanded. On the seventh year the principles of the Sabbath were extended to the land during which it is forbidden to plant, cultivate, or harvest grant, fruit, or vegetables; in the sixth year it is forbidden to plant in order to harvest during the seventh year. Crops that grow untended are not to be harvested by the landlord and are to be left ownerless for all to share, including the poor and animals. On the seventh year, debts contracted by fellow countrymen are to be remitted, which provided temporary relief from these obligations. Judaic social justice is intrinsically ecological justice and vice versa. Contemporary environmentalists have coined the term Eco-Kosher to capture the ethics of care and responsibility (Waskow 1996). Today Jewish environmentalists seek to revive the practice of *Shemitah* not only to express their Eco-Kosher mentality but to create a utopian society (Krantz 2016).

The causal relationship between human conduct and the thriving of the natural environment expresses the relational nature of the covenant: all creatures, both humans and nonhumans, are embedded and interdependent; all exist in a web of relationship with each other and with God; humans who are created in the »image of God« are responsible to God and their task is to care of creation. The quality of the covenant between God and Israel is expressed through the prism of the Land of Israel: when Israel conducts itself according to divine commands, the land is abundant and fertile, benefiting its human inhabitants with the basic necessities of life, but when Israel transgress divine commands, the blessedness of the land is temporarily removed and the land becomes desolate and inhospitable (Lev. 26:32). When the alienation from God becomes so egregious and injustice fills up God's land, God brings about Israel's removal from the land by allowing Israel's enemies to overcome it.

Rabbinic Judaism (70 CE-600 CE) evolved during the exile of Israel from its ancestral land. The rabbis expanded the ecological concerns of the Bible but also gave rise to a scholastic culture that distanced Jews from the natural world. An example of rabbinic extension is the interpretation of the biblical prohibition on destruction of fruit bearing trees in time of siege (Deut. 20:19) to justify the prohibition on all sorts of destruction. The rabbinic principle »Do not destroy« *(bal tashchit)* covered the prohibition on cutting off water supplies to trees, overgrazing; unjustifiably killing animals or feeding them noxious food; hunting animals for sports; species extinction and the destruction of cultivated plant varieties; pollution of air and water; overconsumption of anything and squandering mineral and other resources (E. Schwartz 2001). These environmental regulations indicate that the rabbinic tradition required that one carefully weigh up the ramifications for every interaction with the natural world; it also sets priorities and considers conflicting interests and permanent modification of the environment. Similarly, on the basis of Deut. 22:6 the rabbis articulated the general principle of *tza'ar ba'aley hayyim* which prohibits the affliction of needless suffering of animals. Although rabbinic ethics is undoubtedly hierarchical and human centered—for example, cruelty to animals is forbidden because it leads to cruelty toward humans—the rabbis often presented animals as moral exemplars and recognised special animals as »animals of the righteous«, who live in perfect harmony with their Creator (Rosenberg 2002; R. Schwartz 2012).

In rabbinic Judaism, duties and virtues are closely intertwined: the observance of divine commands generates the desirable character that behaves in the proper way toward other human beings, and ultimately toward God (Tirosh-Samuelson 2003). The rabbis extolled certain virtues (e.g., humility, modesty, moderation, self-control, generosity and benevolence) and they denounced certain vices (e.g., arrogance, greed, or profligacy) and in the Middle Ages, Jewish philosophers explicated rabbinic virtue ethics in the framework of Aristotelian virtue ethics (Weiss 1991). Rabbinic virtue ethics is compatible with environmentalism: the character traits of the ideal Jew can be applied to protection and conservation of the created world. That awareness, however, is quite recent and it manifests an environmental sensibility that did not exist in traditional Judaism (Sokol 2002). Until the rise of Jewish environmentalism in the 1970s, Jews did not see the connection between rabbinic virtue ethics and the natural environment. Rabbinic Judaism established a culture of learning that glorified the study of Torah above all the other commandments. In this regard, the ideal rabbinic Jew was, as Steven Schwarzschild put it, an »unnatural Jew« (Schwarzschild 1984 [2001]), namely a person whose life was devoted to the observance of divine commands and whose spiritual goal was to transcend human embodied physicality to attain communion with God (in Hebrew: *devekut*). The moral imperative that guided traditional Jews was not derived from observation of nature, a world governed by predation and violence, but from an ethical imperative that transcends the natural order because it cares for the weak and the socially marginal (Wychogrod 1991 [2001]).

Modernity profoundly challenged the traditional worldview and lifestyle. When Jews were granted civil rights in the 19th century, most of them sought integration in Western society and culture but in return Jews were expected to modernise Judaism by differentiating between their »religious« private sphere and the »secular« public sphere (Batnitzky 2013). In the 19th century Jews subjected their tradition to critical examination in light of the Enlightenment's ideals, and one aspect of the drive toward modernization was a new attitude toward the natural world. Nature was no longer that which should be transcended but rather the source of renewal and vitalization of Jewish life in the »secular« world (Finer 2010). In the 19th and early 20th century new forms of Jewish secularism emerged, but all of them had a residue of their religious tradition in which the secular and the religious were never apart (Biale 2015). The most revolutionary of Jewish responses to modernity was Zionism which called on Jews to return to the ancestral Land of Israel where a new, modern, secularised Hebraic culture was to be established and where Jews could regain their political sovereignty. From the start Zionism was conceived as the return of Jews to nature, that is, to embodied physicality that would give birth to a new »muscular Jew« who will be rooted in the land and who will redeem him/herself through physical labor and land cultivation (Neumann 2011).

For the past four decades in the Diaspora and in the State of Israel, Jewish environmentalism has emerged as a distinctive voice within contemporary Judaism. In the Diaspora, where Jews are a tiny religious minority, a grass-root Jewish environmental movement educates Jews about environmental matters, inspires Jews to lead an environmentally correct lifestyle, implements »green« communal practices, and rallies Jews to support environmental legislation through interfaith advocacy. The main activities of Jewish environmental organizations and initiatives include nature education, environmental awareness, advocacy on environmental legislation and community building. Jewish environmental advocacy has transformed the practices of Jewish institutions, be they synagogues, day schools, communal organizations, Jewish community centers, and youth movements. Today there are many organizations, programs, and initiatives that promote sustainable practices (e.g., energy efficiency, elimination of plastics, recycling, and waste reduction programs); reduce consumption and promote new eating habits; plant community gardens; link sustainable agriculture to urban Jewish life and education; include environmental issues in the education of youngsters and adults, organise nature walks and outdoor activities; celebrate Jewish holidays (especially Sukkot, Shavuot and Tu Bishvat) with attention to environmental agriculture themselves; promote justice in food production with attention to sustainable agriculture and compassionate treatment of farm animals, and encourage Jews to live sustainably. These programs transcend congregational and denominational boundaries and are often carried out in inter-faith settings in collaboration with non-Jewish organizations.

In Israel, where the Jews are the majority, Zionism has endowed the physical environment of the Land of Israel, its topography, flora and fauna, with spiritual (albeit secular) significance, inculcating intimate knowledge of the Land through nature hikes, field trips and camping. Paradoxically, the outdoor culture has enabled secular Israelis to understand the natural imagery and metaphors of the Bible, the document that legitimised the Zionist nationalist project, while at the same time distance themselves entirely from rabbinic and post-rabbinic Judaism. Jewish environmentalism in Israel is thus overwhelmingly secular. More problematically, the success of the Zionist project exacted a toll on the fragile environment of the Land of Israel:

steep rise in population, rapid urbanization, the on-going Arab-Israeli conflict and initial mistakes about resource management have generated a long list of environmental problems (e. g., air and water pollution, soil erosion, overuse of water, etc.) requiring legislative solutions. Today the state of Israel addresses these environmental challenges through a mixture of policies, legislation, and alternative technologies (e. g., hydropic agriculture, solarization and de-salinization) and environmentalism thrives in Israel through green political parties, numerous environmental NGOs, and creative education and trainings programs (Tal 2002; Orenstein, Tal, and Miller 2013). Many of these environmental initiatives and organizations deal with concrete environmental problems without reference to Judaism, but some organizations draw direct inspiration from Jewish religious sources in their theoretical justification and educational programs. The degree to which Israeli environmentalism should be grounded in traditional Jewish sources is hotly debated in Israel and the movement is quite different from its American counterpart.

Particularly in America, Jews have generated eco-theology which has sought to »re-enchant« nature, namely restore the symbolic meaning of nature so as to integrate environmentalism with the religious sources of Judaism. The re-enchantment of nature has taken two routes: one inspired by Hasidism and Kabbalah and the other inspired by ancient paganism and Goddess religion. The former trend was launched by Schachter-Shalomi (1993; 2007) who was brought up in Hasidism but who called for a »paradigm shift« in Judaism that will put an end to ecocide. Schachter-Shalomi inspired others, especially in the Jewish Renewal Movement, to articulate a Neo-Hasidic eco-theology that espouses panentheism and highlights the presence of God in all aspects of reality (Green 2002; 2010; Seidenberg 2015). A somewhat different focus is evident in the attempts to articulate Earth-based Judaism rooted in ancient Israelite rituals which had been repudiated by the Deuteronomic reform of the 7th century BCE (Golden 2015).[1] Other Judaic attempts to reenchant nature are inspired by Shamanism, Eastern mysticism, Feng Shui, or feminist Earth-based spirituality (Winkler 1998; 2003), giving Jewish environmental spirituality a distinct eclectic flavor common to progressive spirituality (Lynch 2007). Regardless of the sources of inspiration, the Jewish environmental movement in America offers an environmentally conscious way of being Jewish in the world, a way that is particularly appealing to millennial Jews or to Jews who have been previously alienated from the organised Jewish community.

3 Judaism, Ethics of Care, Ecofeminism, and Environmental Virtue Ethics

The ethics of care and responsibility is also shared by Christian and Muslim environmentalism. In Christian environmentalism, the ethics of care and responsibility comes under the banner of either »stewardship« or »creation care« (Ball et al 1992; Northcott 1996; Rasmussen 1996; Kearns 1996; Hessell and Reuther 2000; Moo and Moo 2018) as Christian eco-theologians have translated environmental concerns into religious communication, practices, and rituals. Islam, in which the human being is believed to be a vice regent (Khalifa) of creation, also sees the task of humanity as managing the balance and harmony of the created world and gave rise to eco-theology and environmental activism (Foltz, Denny, Baharuddin 2003; Foltz 2005). Stewardship of the Earth is similarly promoted by secular environmental ethicists who see human beings not as the owners but rather as the care takers of the Earth (Attfield 1983; Callicott 1994). In other words, the discourse on stewardship or care of the Earth is intercultural, interdisciplinary, interreligious, interdenominational, and international (Rozzi 2015). However, religious environmentalists who promote creation care and many secular environmentalists who advocate for Earth stewardship have made relatively little use of ethical theories that put the practice of care at the front and center of ethics. I refer to the feminist ethics of care, a distinctive moral theory that is linked to virtue ethics, on the one hand, and to ecofeminism, on the other hand. Let me clarify how these discourses relate to each other, how they intersect with Judaism, and how they integrate religious and secular dimensions.

The intersection of environmentalism, feminism and ethics originate in the social movements of the 1960s: the Civil Rights movement, the anti-Vietnam War movement, the students' movement, and the women's movement. Despite their difference they all shared the critique of modernity and the assumptions of the Enlightenment Project. At the forefront was feminism which exposed the sexism, inequality and injustice that women suffer in patriarchal society. Since its emergence, the so-called second-wave feminism has transformed all aspects of society (e.g., economy, law, politics, education, arts, and even sports) and profoundly impacted the academy as women demanded inclusion, equality, and justice and promoted gender as category of analysis. Within the discipline of philosophy, feminism has made profound impact as feminist philosophers challenged many assumptions and conventions of the Enlightenment Project about the human Self and the construction of knowledge (Code 1991; Fricker and Hornsby 2000). Within the sub-discipline of ethics, feminist philosophers critiqued the dominant liberal theories: Kantian deontology, consequentialism/utilitarianism, and justice theory. These theories understood society as an aggregate of individuals who are autonomous, independent, and rational, and constructed moral theories in terms of abstract rules and values, be it duty, rights, impartiality, utility, justice, or fairness. By contrast, feminist theorists insisted that the Self is relational, dependent, and vulnerable and that human relations, especially the practice of care, rather than abstract rules constitute the core of ethics.

Carol Gilligan (1982) was the first to articulate the feminist ethics of care when she criticised the work of her mentor, Lawrence Kohlberg, on moral development of the human being. Born Jewish, Gilligan's Jewish identity was rooted in Reconstructionist Judaism, a socially progressive interpretation of Judaism that promoted the holistic view of Judaism as a civilization, integrating »religious« and »secular« dimensions (Kaplan 1934; Goldsmith, Scult and Seltzer 1990). Gilligan attended Hebrew school at the Society for the Advancement of Judaism in New York, which was founded by Rabbi Mordecai M. Kaplan in 1922, and her understanding of the Self reflected the relational, communal, and worldly outlook of Reconstructionist Judaism. Trained in psychology, Gilligan argued that Kohlberg's theory of moral development was wrong because it ignored the different perspectives (or »voices«) of men and women. Whereas men focus on abstract principles of impartial justice and abstract responsibilities, women focus more on people and relationship that emerge within relationship and in concrete situations. Women's moral outlook emphasises solidarity, community, and caring but their moral voice has been silenced by patriarchy. The care perspective has been ignored or trivialised because women were traditionally in positions of limited power and influence.

Gilligan's path-breaking studies gave rise to the feminist ethics of care that construed the Self as relational, interdependent, vulnerable, and concerned with the wellbeing of others. The Self is always in relationship and needs to be understood from this perspective. The feminists, interestingly enough, were not the first to articulate a relational view of the Self. Male Jewish philosophers, most famously Martin Buber in his *I and Thou* (1922), already conceptualised the Self relationally when he distinguished between two types of primary relations: I-It and I-Thou. The former is conditional, instrumental, mediated, treating the Other as a means to an end, whereas the latter is unconditional, non-instrumental, and immediate, treating the Other as an end to whom the Self is intrinsically responsible. The dialogical philosophy of Buber and other Jewish »existentialist« philosophers (e.g., Franz Rosenzweig and Emmanuel Levinas), does not mean that Judaism is inherently feminist or that feminist ethics of care is inherently Jewish, or that Jewish dialogical philosophers are necessarily concerned about the environment. Indeed, Buber was no feminist, Rosenzweig and Levinas, who used the category of »the feminine« extensively in their philosophy, had no interest in ecology, and Rosenzweig's scathing critique of paganism could be applied to environmentalism (Tirosh-Samuelson 2005: 289). However, feminist ethics of care has much in common with Jewish dialogical philosophy because both highlight the relational nature of the Self, the focus on vulnerability and dependency, and the importance of responsibility and responsiveness (Batnizky 2004; Diedrich, Burggraeve, and Gastmans 2006; C. Taylor 2005). Also the ideas of Buber and Levinas have already been applied to environmentalism (McFague 1997; Edelgrass, Hatley and Diehm 2012): for Buber nature is a moral subject (Buber 1970 [1923]:75) and for Levinas nature is the Other that places infinite responsibility on the Self (Levinas 1969). Jewish dialogical philosophers and feminist ethicists share the focus on relationality, responsibility, and care.

Care, the practice in which we attend to the needs of others, characterises the relational Self. In patriarchal society caring is gendered: for the duration of their life women are the primary care givers, caring for children, sick spouses, and aging parents. Feminist theorists who promoted ethics of care as alternative theory originally argued that caring manifests a distinctive female morality. For example, Nel Noddings who argued that caring is »rooted in receptivity, relatedness, and responsiveness« (Noddings 1984: 2) closely analyzed the dynamic relationship between the care giver and care recipient although to whom care is due remained a debated issue. Sara Ruddick

(1989) used care to theorise the experience of mothering which she viewed as a unique approach to moral reasoning and a ground for feminist politics of peace. Virginia Held (1995; 2006) construed care as the most basic core value and explored the characteristics of the caring personality; in so doing she linked ethics of care to virtue ethics, while insisting on the differences between them. And Eva Feder Kittay (2011) developed a dependency-based account of equality rooted in the activity of caring for the seriously disabled and called for institutional reforms of professional care workers, recommending welfare for all care givers, akin to worker's compensation or unemployment benefits.

The ethics of care, however, was also critiqued from within the feminist discourse by Joan Tronto, a Jewish woman whose upbringing in New York was not so different from Carol Gilligan's. Tronto (1993) chided feminists for associating care exclusively with women and for disregarding the political context within moral arguments. Trained at the City University of New York, the bastion of Jewish academic secularism in the 1960s and 1970s, Tronto focused on the political implications of the practice of care.[2] Care, she argued, has always been undervalued and disregarded in order to uphold the structures of power and privilege. Rejecting the notion that any work connected to care ought to be done women, or the notion of feminine morality, Tronto argued that everyone as citizen of a democracy has a responsibility towards care. Toronto called for the moral boundaries to be shifted in order to create more caring societies. In her *Caring Democracy* (2013) Tronto elaborates the argument: if everyone accepts the responsibility and participates in allocation of care responsibility, some fundamental values and commitment will be addressed. Care should be included as a public concern in order to strive for true freedom, equality and justice for all citizens. The key elements of care—attentiveness, responsibility, competence, and responsiveness—should be understood politically and not only ethically. Tronto's political emphasis would deeply impact the feminist ethics of care as well as ecofeminism.

Feminist ethics of care was not the only moral theory to challenge the reigning ethical theories of Kantian deontology and consequentialism. Another approach to ethics—virtue ethics—did the same without reference to feminism, although the people who argued for it were female British philosophers—G. E. M. Anscombe, Philippa Foot, and Rosalind Hursthouse. Already in the late 1950s, Anscombe (1958) charged that secular approaches to moral theory are without foundation because they use concepts such as ›ought‹, ›obligated‹ and ›right‹ that are all legalistic and require a legislator as the source of moral authority. In the past God occupied that role, but secular moral theories that dispense with God as part of the theory are lacking the proper foundation for meaningful employment of these concepts. As an alternative she insisted that we need to develop a moral theory based on moral psychology, moral facts, the facts of human nature, and an account of the good for humans. A similar critique came from Philippa Foot who like Anscombe urged philosophers to take their inspiration not from Kant or Mill but from Plato, Aristotle, and the Stoics. The result was the emergence of virtue ethics, which focused on character, virtue, and human flourishing as an alternative moral theory to the reigning approaches that focused on duties, abstract rules, or the consequences of actions. In virtue ethics the question what should one do is inseparable from the question what kind of person one should be (Statman 1997; Slote 1992; Crisp and Slote 1997; Hursthouse 1999). The precise relationship between ethics of care and virtue ethics continues to be debated: some have argued that ethics of care cannot and should not constitute a comprehensive moral theory, others have claimed that ethics of care has its own moral domain, for example, friendship, and still other have preferred to view ethics of care as part of the more comprehensive moral framework, namely virtue ethics (Halwani 2003).

As feminist ethics of care and virtue ethics came into their own, so did environmental ethics, the philosophical response to the environmental crisis (Jamieson 2001). Traditionally, ethics was viewed as distinctly human: only humans are moral agents and the ethical situation is restricted to social relations among human beings. The environmental crisis brought new questions to the fore: Do humans have obligations to the natural world? Does nature have a moral standing? In Kantian deontology »moral duties to nonhumans exist, but they are secondary to the primary imperative to treat rational human beings as ends in themselves« (Keller 2010:10). Environmental ethicists challenged this anthropocentric perspective by posing the question: »Do human beings really constitute the entire scope of moral considerability?« At first, environmental ethicists argued that standard moral theories are sufficient to ground environmental ethics by extending moral considerability to non-humans (Singer 1975; Regan 1982) but most environmental ethicists went beyond »extensionism«

to argue that the moral categories of Western ethics cannot ground environmental ethics: a new paradigm rooted in non-individualistic ontology is needed, which can take into consideration ecological wholes, inclusive of human and nonhuman biota and the abiotic environment. By including non-humans in the moral situation, environmental ethics moved beyond the anthropocentric perspective of traditional moral philosophy which identified humans exclusively as the subject matter of ethics. Like feminism, environmentalism challenged existing moral theories and gave rise to various strands that debate the ontological basis of environmental ethics: »To whom is care due, why, and how should care be given«?

The theoretical debates among ethicists crystalised in two major approaches—Deep Ecology and Social Ecology—although these two do not exhaust the range of environmental moral theories (Krebs 1999). The former was non-anthropocentric (also known as »biocentric« or »ecocentric«) whereas the latter was anthropocentric. Deep Ecology (Naess 1973; Devall and Sessions 1985; Sessions 1995) articulated egalitarian biocentrism and metaphysical holism. Dismissing conservation and resource management as »shallow ecology«, Deep Ecology advocated deep rethinking of the place of humanity in the ecological whole. Deep Ecology was concerned with the flourishing of non-human life, values and diversity of life and the need to sustain the very conditions for the diversity of myriad forms of life. By contrast, Social Ecology claimed that the source of the ecological crisis is the very human tendency for domination. Murray Bookchin (1990), a Jewish left-leaning social philosopher, acknowledged that humans are part of nature but also insisted that humans are also more than just nature. As a Marxist critic of Darwinism, Bookchin argued that nature developed not through competition but through cooperation and mutualism to every greater possibility for diversity, freedom and subjectivity. Humans are indeed part of nature but they are also a »quantum leap« within the natural process: only humans actualise the potentiality for nature to become self-conscious and free and therefore only they have the responsibility to nature, contrary to the instrumentalism of modern industrial capitalism and in contrast to biocentric egalitarian approach of Deep ecology. Human responsibility to nature could be properly carried out only if humans first eliminate practices of exploitation, domination and hierarchy by developing communitarianism.

Elimination of oppression was also the focus of ecological feminists, or ecofeminists.[3] Ecofeminists brought to the fore the gender dimension of the environmental crisis: in all cultures and in all societies women have been associated with »nature« so that »the oppression and exploitation of women and the oppression and exploitation of nature are intimately connected and mutually reinforcing« (Warren 1996). Therefore, it is impossible to liberate nature or nonhuman animals without taking the oppression of women seriously, or liberate women without taking the oppression of nature (or specifically nonhuman animals) seriously. Ecofeminists thus sought to make feminism environmental and conversely reconceive environmental ethics in feminist categories (Gaard 1993; Mies and Shiva 2014 [1993]; Adams and Gruen 2014). The essentialists among them argued that women are closer to nature and the ethics of care expresses »the essential voice of women's lived experience and sense of self as embedded in relationship« (King 1991: 76). Other ecofeminists took a conceptualist approach arguing that »an ecofeminist environmental ethics must redress the conceptual opposition that patriarchal culture sets up between men and culture on the one hand and women and nature on the other. These dualisms underlie a parallel that exists between the oppression of women and the domination of nature in a patriarchal society« (King 1991: 76). Regardless of these differences, all ecofeminists decry male domination that supports the oppressive framework of patriarchy, but they are deeply divided about how to accomplish the egalitarian goal, how to justify it philosophically or how to relate ecofeminism to religion. Some secular ecofeminists align themselves with Social Ecology (e.g., Biehl 1991; Salleh 1997) while others promote Earth-based spirituality and Goddess religion that are more attuned with Deep Ecology (Reuther 1992; Christ 1998; 2003), and still others (Plumwood 1993, 2002) argued that ecofeminism should focus on the development of a particular character that could care for the environment.

Generally speaking, ecofeminism offers a feminist orientation to Earth care (Merchant 1995 [1991]), but ecofeminism is too variegated and nuanced to be summarised here. Philosophically speaking, ecofeminism is contextual, pluralistic, concrete, and situational, making no attempt to be »unbiased«, »value-neutral« or »objective« (Warren 1994; 1996). Speaking from the standpoint of the oppressed (be they women or nature), feminist ecological ethics rejects as meaningless or untenable any gender-free or gender-neutral description of humans, but how the ethics of care relates to ecofeminism is a matter of dispute. Some ecofeminists promote the integration of feminist care ethics and environmental ethics and claim that »feminist environmental ethics of care is not limited to caring for sentient beings« but should be extended to all existents (Whyte and Cuomo 2017). Furthermore, women take a particular role in environmental protection because of their local knowledge about communal wellbeing, »especially where basic rights and needs are threatened by destructive projects that generate profits for outsiders« (Whyte and Cuomo 2017: 243). In the age of climate change and global warming, such local knowledge is most crucial for survival, making the social caring and ecological care-taking interwoven. Yet, there are other ethicists who have noted that while ethics of care has »an intuitive appeal« to feminist ecological ethics »without further development into a political dimension, Gilligan's research may be turned against feminist and ecofeminist objectives« (Curtin 1991: 66). Indeed, recent ecofeminist discourse (e.g., Sandilands 1999; Cuomo 1998; Macgregor 2006; Kheel 2008) follows the lead of Joan Tronto by contextualising and politicising the practice of care and inquiring who does the caring and who or what is cared for; who gets to make these decisions; what models of human-to-human care are we invoking in the process, and what are the gender dynamics of our models of care.

The more ecofeminism concerned itself with the practice of care and the function of women in actual communities, the more it had to pay attention to religious traditions, since religion frames the cultures of traditional societies.[4] Secular ecofeminists have acknowledged that caring toward nonhuman others »is a basic good in African traditions«, that ethics of care has parallels in Buddhism, »where there is a foundational commitment to compassion for all sentient being«, and that there is considerable overlap between »feminist and indigenous conceptions of care ethics« (Whyte and Cuomo 2017: 243). The relevance of Jews and Judaism to ecofeminism and feminist ethics of care, however, has often been overlooked. That is unfortunate because Jewish women (e.g., Betty Friedan, Bella Abzug, Letty Cottin Pogrebin among many others) were at the leadership of the feminist movement and within ecofeminism many of the proponents of Earth-based spirituality were born Jews, whose feminist critique of patriarchal Judaism led them to adopt ecofeminism as a spiritual alternative to traditional Judaism, without denying their Jewish (ethnic) identity. Thus Starhawk (1979) (aka Miriam Simos) became the High Priestess of the Wicca cult in the United States, Riane Eisler (1989), a Jewish human rights activist, highlighted egalitarian and universalist dimension of ecofeminism; and Irene Diamond and Gloria Orenstein (1990) teased out the political implications of ecofeminism (Tirosh-Samuelson 2012). For all of these Jewish-born ecofeminists Earth-care is a spiritual practice and women are the paradigmatic care takers.

Feminist ethics of care, virtue ethics, and ecofeminism all focus on the moral quality of the agent rather than on abstract rules or on the consequences of the act. But the arguments for integrating virtue ethics and environmentalism were articulated not by ecofeminists but by environmental ethicists who promoted environmental virtue ethics (van Wensveen 2000; Sandler and Cafaro 2005; Treanor 2014). Since the First Earth Day in 1970 the environmental movement scored many successes, but in the beginning of the 21st century it became clear that environmentalism has failed to change human behavior toward the environment. Global warming, climate change, the destruction of marine life, extreme weather events, desertification and pollution of air, water, and soil have all worsened notwithstanding important legislative achievements. This awareness led Ronald Sandler (2007) to insist that environmentalism has to move from the focus on legislative activity to the cultivation of character, since »a virtue-oriented approach provides action guidance in environmental contexts and on environmentally related issues« (Sandler 2007: 102). Environmental virtue ethics insisted on the need to cultivate an environmental personality type that is disposed to appreciate, respect, wonder and love nature. The emergence of the environmentally good personality should be the focus of education. The advocates of environmental virtue ethics maintain that environmental virtues are not limited to character traits that enhance our experience in environmental contexts (e.g., openness, appreciation, receptivity, love and wonder) but include as well character traits such as temperance, fortitude, commitment, optimism and cooperation that are favorable dispositions to securing environmental goods, resources and opportunities. By contrast, environmental vices are those dispositions that are detrimental to environmental health at the level needed to provide the good necessary for humans to flourish, or traits that prevent us from realising benefits that the natural environment can provide. Environmental virtue ethics is politically relevant because action is right to the extent that it best accomplishes the target of the operative virtues for a particular agent in a particular situation. The virtues provide the standard of rightness; action guidance is accomplished through their application to a concrete situation.

Interestingly, and perhaps even ironically, although virtue ethics was initially proposed as a secular alternative to Kantian deontology, secular environmental virtue ethics reminds us of the importance of religious values and attitudes. Many environmental virtues have religious analogues, and conversely, virtues praised by the religious traditions are conducive to environmental conduct. Louke van Wensveen, a major contributor to environmental virtue ethics, has correctly observed the ubiquity of virtue language within the environmental discourse: »ecological virtue language turns up in the writings of social ecologists as well as Deep ecologists, bioregionalists as well as animal rights activists, creation theologians as well as environmental philosophers, main stream theologians as well as radical ecofeminists« (van Wensveen 2005: 16). Her list of environmental virtues consists of the following: »adaptability, benevolence, care, compassion, solidarity, connectedness, creativity, cooperation, fostering, friendship, frugality, gratitude, healing, hope, inclusivity, joy, justice, moderation, or restraints, openness, passion, perseverance, realism, self-examination, sensuousness, sharing, spontaneity, vulnerability, wisdom and wonder« (van Wensveen 2005: 21).[5] Needless to say, many of these virtues, values, and attitudes are promoted by world religions as well as in the writings of ecofeminists or feminist ethicists of care. By the same token, the environmental vices she lists—»anthropocentrism, arrogance, carelessness, competitiveness, consumerism or greed, contempt, cruelty, denial, despair, domination or mastery, dualistic thinking, elitism, exploitation, ignorance or thoughtlessness, indifference, insensitivity, manipulation, pride, otherworldliness, reductionism, romanticism, and wastefulness« (van Wensveen 2005: 21)—have parallels in religious traditions although they can also be accused of promoting some of these vices. The relationship between religious virtue ethics and secular environmental virtue ethics requires further discussion, but van Wensveen helps us to see their connection by noting that the person most responsible for »the use of virtue language among ecologically minded people« was no other than Lynn White Jr (van Wensveen 2005: 19)! When he critiqued the Judeo-Christian tradition for authorising domination of nature, he did not intend believers to leave behind their religious commitment but rather to identify the relevant resources within the Judeo-Christian tradition that support a non-exploitative attitude toward nature. For White it was

St. Francis's »belief in the virtue of humility« (White 1967: 1206) that anchors ecological consciousness and caring.

The virtue of humility, of course, is also the primary virtue of rabbinic Judaism (Nelson 1985) in which the paradigmatic humble person was the prophet and lawgiver, Moses. In rabbinic Judaism the virtues are cultivated by following divine commands and living the life of Torah. For the rabbis there is no conflict between ethics of duty and ethics of virtue: rather, it is divine law that enables individuals to cultivate the appropriate virtues in inter-personal relations and in their relationship with God (Tirosh-Samuelson 2003). The virtues extolled by Judaism, such as humility *(tzniut)*, modesty *(anavah)*, continence or self-control *(kibbush ha-yetzer)*, moderation *(metinut)*, and simplicity *(histapkut be-muat)* are all conducive to environmental behavior. The ideal Jew is not one who exploits and plunders nature but rather one who is compassionate, caring, devoted to the well-being of others, including non-human animals. The compassionate treatment of animals is a principle that has guided Jewish religious jurisprudence for centuries as much as it has inspired contemporary laws of the Israeli government that regulate all aspects of animal treatment (1994). The ethics of care and responsibility guide the normative Jewish tradition as well as Jewish meta-ethics as articulated by dialogical philosophers such as Martin Buber, Hans Jonas and Emmanuel Levinas (Werner 2008; Tirosh-Samuelson 2017: 186–194). Jewish dialogical philosophy shares with ecofeminism the relational concept of the self, the attention to community over individualism, the focus on character and virtue cultivation, and the ethics of responsibility and care. The relationship between ecofeminism and Judaism is quite complex: certain aspects of ecofeminism are more compatible with Judaism than others, and much depends on how one interprets Judaism (Tirosh-Samuelson 2005).

4 Conclusion

The severity of the eco-crisis compels us to develop an integrative approach that cuts across boundaries between individuals, nations, religions, philosophies, ideologies, and disciplines. Awareness of the eco-crisis emerged in the 1960s as part of the wholesale examination of the Enlightenment Project and its ideology of »progress« The Holocaust, Hiroshima and the anthropogenic eco-crisis made it clear that humanity is not marching toward a more just, peaceful and verdant world, as the Enlightenment had promised, but rather toward potential destruction of life on Earth, including human life. A plethora of voices and perspectives (e.g., feminism, environmentalism, postcolonialism, posthumanism, postsecularism and others) critiqued the negative aspects of modernity worldview, showing it to be philosophically inadequate, socially unjust, or culturally repressive. In the second half of the 20th century these criticisms generated numerous academic discourses, disciplines, sub-disciplines that flourished side by side in the professional academy with little interaction with each other. Today it is clear that what is needed to combat the destruction of life on Earth is no more theoretical clarification, philosophical refinement, or ideological purism; instead, environmental activism is required and it calls for collaboration and cooperation between international and national bodies, governmental and non-governmental organizations, scientific and humanistic disciplines, religious and secular cultures. This essay has proposed the ethics of care and responsibility as the overarching principle that can give coherence to this multivalent effort because the ethics of care and responsibility bridges theory and praxis, individuals and communities, human and nonhuman worlds, values and actions, and most importantly the secular and the religious.[6] In the Anthropocene caring for the world is a religious obligation as much as it is a secular ethical imperative.

—Hava Tirosh-Samuelson, Arizona State University, USA

References

ed. Adams, Carol J. and Lori Gruen, *Ecofeminism: Feminist Intersections with Other Animals & Earth*, London, 2014.

Angus, Ian, *Facing the Anthropocene: Fossil Capitalism and the Crisis of Earth Systems*, New York, 2016.

Anker, Peter, *Imperial Ecology: Environmental Order in the British Empire, 1895–1945*, Cambridge MA, 2001.

Anscombe, G. E. M., »Modern Moral Philosophy«, *Philosophy*, Vol. 33, No. 124, 1958, pp. 1–19.

Attfield, Robin, *The Ethics of Environmental Concern*, New York, 1983.

ed. Ball, Ian M. Goodall and C. Palmer, *The Earth Beneath: A Critical Guide to Green Theology*, London, 1992.

ed. Baharuddin, Azizan Haji, Frederick Denny and Richard Foltz, *Islam and Ecology: A Bestowed Trust,* Cambridge, MA, 2003.

Batnitzky, Liora Faye, »Dependency and Vulnerability: Jewish and Feminist Existentialist Construction of the Human«, in: *Women and Gender in Jewish Philosophy,* ed. Hava Tirosh-Samuelson, Bloomington, 2004, pp. 127–152.

Batnitzky, Liora Faye, *How Judaism Became a Religion: Introduction to Modern Jewish Thought.* Princeton, 2013.

ed. Berry, Robert James, *Environmental Stewardship: Critical Perspectives, Past and Present,* London and New York, 2006.

Biale, David, *Not in the Heavens: The Tradition of Jewish Secular Thought,* Princeton, NJ, 2015.

Biehl, Janet, *Rethinking Eco-Feminist Politics,* Boston, 1991.

Bonnheuil, Christophe and Jean Baptiste Fressoz, *The Shock of the Anthropocene,* London and New York, 2017.

Bookchin, Murray, *The Philosophy of Social Ecology: Essays on Dialogical Naturalism,* Montreal and New York, 1990.

Bruelle, Robert J. and Riley E. Dunlap. 2015. »Sociology and Global Climate Change: Introduction.« In *Climate Change and Society: Sociological Perspectives,* ed. Riley E. Dunlap and Robert J. Bruulle. Oxford: Oxford University Press. 1–31

Buber, Martin, *I and Thou,* New York, 1970 [1923].

Callicott, J. Baird, *In Defense of the Land Ethics: Essays in Environmental Philosophy,* Albany, 1994 [1989].

Christ, Carol P., *Rebirth of the Goddess: Finding Meaning in Feminist Spirituality,* Reading, 1998.

Christ, Carol P., *She Who Changes: Re-Imagining the Divine in the World,* New York, 2003.

Clingerman, Forrest, and O'Brien, Kevin J., »Playing God: Why Religion Belongs in the Climate Change Debate«, *Bulletin of the Atomic Scientists,* Vol. 70 (3), 2014, pp. 27–37.

Code, Lorraine, *What Can She Know? Feminist Theory and the Construction of Knowledge,* Ithaca, 1991.

ed. Crisp, Robert and Michael Slote, *Virtue Ethics,* Oxford, 1997.

Curtin, Deane, »Toward an Ecological Ethics of Care«, *Hypatia,* Vol. 6, No. 1, 1991, pp. 60–74 (Reprinted in *Ecological Feminist Philosophies,* ed. Karen J. Warren, Bloomington, 1996, pp. 6–80).

Cuomo, Chris J., *Feminism and Ecological Communities: An Ethics of Flourishing,* London and New York, 1998.

ed. Diamond, Irene and Gloria Feman Orenstein, *Reweaving the World: The Emergence of Ecofeminism,* San Francisco, 1990.

Dalby, Simon, »Framing the Anthropocene: The Good, the Bad and the Ugly«, *The Anthropocene Review,* Vol. 3 (1), 2016, pp. 31–51.

ed. Deanne-Drummond, Celia, Sigurd Bergmann, and Markus Vogt, *Religion in the Anthropocene.* Eugene, Oregon, 2017.

ed. Devall, Bill and George Sessions, *Deep Ecology: Living as If Nature Mattered,* Salt Lake City, 1985.

Diedrich, Wolf W., Roger Bruggraeve, and Chris Gastmans, »Toward a Levinasian Care Ethics«, *Ethical Perspectives*, Vol. 13, No. 1, 2006, 32–59.

ed. Edelglass, William, James Hatley and Christian Diehm, *Facing Nature: Levinas and the Environment*, Pittsburgh, 2012.

Eisler, Riane Tenenhaus, *The Chalice and the Blade: Our History, Our Future*, New York, 1989.

Elis, Erle C, *The Anthropocene: A Very Short Introduction*, Oxford, 2018.

Finer, Shmuel, *The Origins of Jewish Secularization in Eighteenth-Century Europe*, trans. Chaya Naor. Philadelphia, 2010.

ed. Foltz, Richard C., *Environmentalism in the Muslim World*, New York, 2005.

ed. Fricker, Miranda and Jennifer Hornsby, *The Cambridge Companion to Feminism in Philosophy*, Cambridge, UK, 2000.

ed. Gaard, Greta, *Ecofeminism: Women, Animals, Nature*, Philadelphia, 1993.

Gilligan, Carol, *In Different Voice: Psychological Theory and Women's Development*, Cambridge MA, 1982.

Golden, Zelig, *Earth Based Judaism: Early Origins, Polemics and a New Vision*, Graduate Theological Union, 2015 (MA Thesis).

ed. Goldsmith, Emanuel S., Mel Scult, and Robert M. Seltzer, *The American Judaism of Mordecai M. Kaplan*, New York, 1990.

Gottlieb, Roger S., *The Oxford Handbook of Religion and Ecology*, Oxford, UK, 2006.

Green, Arthur, »A Kabbalah for the Environmental Age«, in: *Judaism and Ecology: Created World and Revealed Word*, ed. Hava Tirosh-Samuelson, Cambridge, MA, 2002, pp. 3–15.

Green, Arthur, *Radical Judaism: Rethinking God and Tradition*, New Haven, 2010.

Grim John, and Tucker, Mary Evelyn, *Ecology and Religion*, Washington DC, 2014.

ed. Habel, Norman C., *Readings from the Perspective of the Earth*, The Earth Bible 1. Sheffield, 2000.

Halwani, Raja, »Care Ethics and Virtue Ethics«, *Hypatia*, Vol. 18, No. 3, 2003, pp. 161–190.

Hargrove, Eugene C, *Religion and the Environmental Crisis*, Athens, 1986.

ed. Hart, John, *The Wiley-Blackwell Companion of Religion and Ecology*. Malden MA and Oxford UK, 2017.

Hedrick, Daniel, *The Tentacles of Progress: Technology Transfer in the Age of Imperialism, 1850–1940*, Oxford, 1988.

ed. Held, Virginia, *Justice and Care: Essential Readings in Feminist Ethics*, Boulder, 1995.

Held, Virginia, *Ethics of Care: Personal, Political and Global*, Oxford and New York, 2006.

Helfand, Jonathan, »Ecology and the Jewish Tradition«, *Judaism: A Quarterly Journal*, Vol. 20, 1971, pp. 330–335.

Helfand, Jonathan, »The Earth is the Lord's: Judaism and Environmental Ethics«, in: *Religion and Environmental Crisis*, ed. Eugene C. Hargrove, Athens, GA and London, 1986, pp. 38–52.

Henson, Robert, *The Thinking Person's Guide to Climate Change*, Boston, 2014.

ed. Hessel, Dietrich and Rosemary Radford Reuther, *Christianity and Ecology: Seeking the Well-Being of Earth and Humans,* Cambridge, MA, 2000.
Horrell, David G., *The Bible and the Environment: Towards a Critical Ecological Biblical Theology,* London, 2010.
Hulme, Mike, *Why We Disagree about Climate Change: Understanding Controversy, Inaction and Opportunity,* Cambridge, 2009.
Hursthouse, Rosalind, *On Virtue Ethics,* Oxford, 1999.
Hüttermann, Aloys, *The Ecological Message of the Torah: Knowledge, Concepts and Laws Which Made Survival in the Land of »milk and honey« Possible,* Atlanta, 1999.
ed. Jamieson, Dale, *Companion to Environmental Philosophy,* Oxford, 2001.
ed. Jenkins, Willis, Mary Evelyn Tucker and John Grim, *The Routledge Handbook of Religion and Ecology,* New York and London, 2017.
Kaplan, Mordecai, *Judaism as a Civilization,* Philadelphia, 1934.
Kay, Jeanne, »Concepts of Nature in the Hebrew Bible«, *Environmental Ethics,* Vol. 10 (4), 1988 [2001], pp. 309–327 [Reprinted in *Judaism and Environmental Ethics: A Reader,* ed. Martin D. Yaffe, 86–104. Lanham, MD: Lexington Books].
Kearns, Laurel, »Saving the Creation: Christian Environmentalism in the United States«, *Sociology of Religion* Vol. 57, 1996, pp. 55–70.
Kearns, Laurel, »The Role of Religions in Activism«, in: *The Oxford Handbook of Climate Change and Society,* ed. John S. Dryzek, Richard B. Norgaard and David Schlosberg, Oxford, 2011, pp. 414–428.
ed. Keller, David R., *Environmental Ethics: The Big Questions,* Malden MA and Oxford UK, 2010.
Kheel, Marti, *Nature Ethics: An Ecofeminist Perspective,* Lanham, 2008.
King, Robert J. H., »Caring about Nature: Feminist Ethics and the Environment«, *Hypatia,* Vol. 6, No. 1, 1991, pp. 75–89. Reprinted in *Ecological Feminist Philosophies,* ed. Karen J. Warren Bloomington, 1996, pp. 82–96.
Kittay, Eva Feder, »The Ethics of Care, Dependence, and Disability«, *Ratio Juris: An International Journal of Jurisprudence and Philosophy of Law,* Vol. 24, 2011, pp. 49–58.
Krantz, David, »The Shemita Revolution: The Reclamation and Reinvention of the Sabbatical Year«, *Religions,* Vol. 7, No. 8, 2016, pp. 1–31.
Krebs, Angelika, *Ethics of Nature: A Map,* Berlin and New York, 1999.
Lamm, Norman, »Ecology in Jewish Law and Theology«, in: *Faith and Doubt: Studies in Traditional Jewish Thought.* New York, 1972.
Levinas, Emmanuel, *Totality and Infinity: An Essay on Exteriority,* trans. Alphonso Lingis, Pittsburgh, 1969.
ed. Light, Andrew and Eric Katz, *Environmental Pragmatism,* New York and London, 1996.
Liu, J. et al. »Complexity of Coupled Human and Natural Systems«, *Science,* Vol. 317, 2007, pp. 1513–1516.
Lynch, Gordon, *The New Spirituality: An Introduction to Progressive Belief in the Twenty-First Century,* London, 2007.
MacGregor, Sherilyn, *Beyond Mothering Earth: Ecological Citizenship and the Politics of Care,* Vancouver and Toronto, 2006.

McFague, Sally, *Super, Natural Christians: How We Should Love Nature*, Minneapolis, 1997.
McGrath, Alister, *The Reenchantment of Nature: The Denial of Religion and the Ecological Crisis*, London, 2002.
Merchant, Carolyn, *Earthcare: Women and the Environment*, New York, 1995 [1991].
ed. Mies, Maria and Vandana Shiva, *Ecofeminism*, London, 2014 [1993].
Minteer, Ben, *Refounding Environmental Ethics: Pragmatism, Principle, and Practice*, Philadelphia, 2011.
Moo, Douglas J. and Jonathan Moo, *Creation Care: A Biblical Theology of the Natural World*, San Francisco, 2018.
Naess, Arne, *Ecology, Community and Lifestyle: Outline of an Ecosophy*, trans. and rev. David Rothenberg, Cambridge, England and New York, 1973.
Nelson, Daniel M., »The Virtue of Humility in Judaism: A Critique of Rationalist Hermeneutics«, *The Journal of Religious Ethics*, Vol. 13, No. 2, 1985, pp. 298–311.
Neumann, Boaz, *Land and Desire in Early Zionism*, Newton, 2011.
Noddings, Nel, *Caring: A Feminine Approach to Ethics and Moral Education*, Berkeley, 1984.
Noddings, Nel, *Starting at Home: Caring and Social Policy*, Berkeley, 2002.
ed. Noddings, Nel, *Educating Citizens for Global Awareness*. New York, 2005.
Northcott, Michael S., *The Environment and Christian Ethics*, Cambridge, UK, 1996.
Oelschlaeger, Max, *Caring for Creation: An Ecumenical Approach to the Environmental Crisis*, New Haven, 1994.
ed. Orenstein, Daniel, Alon Tal and Char Miller, *Between Ruin and Restoration: An Environmental History of Israel*, Pittsburgh, 2013.
Plumwood, Val, *Feminism and the Mastery of Nature*, London, 1993.
Plumwood, Val, *Environmental Culture: The Ecological Crisis of Reason*, London, 2002.
Rasmussen Larry L., *Earth Community, Earth Ethics*, Maryknoll, 1996.
Rasmussen Larry L., *Earth-Honoring Faith: Religious Ethics in a New Key*, Oxford, UK, 2012.
Regan, Tom, *All that Dwell Therein: Animal Rights and Environmental Ethics*, Berkeley, 1982.
Richter, Sandra, »Environmental Law in Deuteronomy: One Lens on Biblical Theology of Creation Care.« *Bulletin for Biblical Research*, Vol. 20, No. 3, 2010, pp. 355–376.
Rosenberg, Shalom, »Concepts of Torah and Nature in Jewish Thought«, in: *Judaism and Ecology: Created World and Revealed Word*, ed. Hava Tirosh-Samuelson, Cambridge, MA, 2002, 189–226.
ed. Rozzi, Ricardo, et al. *Earth Stewardship: Linking Ecology and Ethics in Theory and Practice*. Heidelberg, Dordrecht, London and New York, 2015.
Ruddick, Sara, *Maternal Thinking: Towards a Politics of Peace*, Boston, 1989.
Ruddiman, William F., »The Anthropocene«, *Annual Review of Earth Planetary Science* Vol. 41, 2013, pp. 45–68.

Ruddiman, William F, Eric C. Ellis, Jed O. Kaplan, and Dorian Q. Fuller. »Defining the Epoch We Live In: Is a Formally Designated ›Anthropocene‹ a Good Idea?« *Science,* Vol. 348, Issue 6230, 2015, pp. 38–39.

Reuther, Rosemary Radford, *Gaia & God: An Ecofeminist Theology of Earth Healing,* San Francisco, 1992.

Salleh, Ariel, *Ecofeminism as Politics: Nature, Marx and the Postmodern,* London and New York, 1997.

Sandilands, Catriona, *The Good-Natured Feminist: Ecofeminism and the Quest for Democracy,* Minneapolis and London, 1999.

Sandler, Ronald L., *Character and Environment: A Virtue-Oriented Approach to Environmental Ethics,* New York, 2007.

Schachter-Shalomi, Zalman, *Paradigm Shift,* ed. Ellen Singer, Northvale, 1993.

Schachter-Shalomi, *Integral Halachah: Transcending and Including Jewish Practice through the Lens of Personal Transformation and Global Consciousness,* ed. Daniel J. Siegel, Victoria BC, 2007.

Schwartz, Eilon, »*Bal Tashchit:* A Jewish Environmental Precept«, *Environmental Ethics,* Vol. 19, No. 4, 1997, pp. 355–374.

Schwartz, Richard H, »Jewish Traditions«, in: *Animals and World Religions,* ed. Lisa Kammerer, New York, 2012, pp. 169–204

Schwarzschild, Steven S., »The Unnatural Jew«, *Environmental Ethics,* Vol. 6, No. 4, 1984, pp. 347–362 (Reprinted in *Judaism and Environmental Ethics: A Reader,* ed. Martin D. Yaffe, Lanham, 2001, pp. 267–282).

Sessions, George, *Deep Ecology for the Twenty-First Century: Readings on the Philosophy and Practice of New Environmentalism,* Boston, 1995.

Seidenberg, David Mevorach, *Kabbalah and Ecology: God's Image in the More-Than-Human World,* Cambridge, 2015.

Singer, Peter, *Animal Liberation,* 2nd ed., New York, 1975.

Slote, Michael, *From Morality to Virtue,* Oxford UK, 1992.

Sokol, Moshe, »What Are the Ethical Implications of Jewish Theological Conceptions of the Natural World?« in: *Judaism and Ecology: Created World and Revealed Word,* Cambridge, MA, 2002, pp. 261–282.

Starhawk, *The Spiral Dance: A Rebirth of the Ancient Religion of the Great Goddess,* San Francisco, 1979.

ed. Statman, Daniel, *Virtue Ethics: A Critical Reader,* Cambridge, 1997.

Steffen, Will, Jacques Grinevald, Paul Crutzen, and John McNeil, »The Anthropocene: Conceptual and Historical Perspectives«, *Philosophical Transactions of the Royal Society* 369, 2011, 842–867.

Tal, Alon, *Pollution in the Holy Land: An Environmental History of Israel,* Los Angeles and Berkeley, 2002.

ed. Taylor, Bron, *Encyclopedia of Religion and Nature,* 2 vols., London, 2004.

Taylor, Chloe, »Levinasian Ethics and Feminist Ethics of Care«, *Symposium: Canadian Journal of Continental Philosophy/Revue canadienne de philosophie continentale,* Vol. 9, No. 2, 2005, pp. 217–239.

Tirosh-Samuelson, Hava, *Happiness in Premodern Judaism: Virtue, Knowledge, and Well-Being.* Cincinnati, 2003.

Tirosh-Samuelson, Hava, »Religion, Ecology and Gender: A Jewish Perspective«, *Feminist Theology,* Vol. 13, No. 3, 2005, pp. 373–397.

Tirosh-Samuelson, Hava, »Jewish Environmentalism: Bridging Scholarship, Faith and Activism«, In *Jewish Thought, Jewish Belief*, ed. Daniel Lasker, English Section, Beer Sheba, 2012, pp. 65–117.

Tirosh-Samuelson, Hava, »Jewish Environmental Ethics: The Imperative of Responsibility«, in: *The Wiley Blackwell Companion to Religion and Ecology*, ed. John Hart, Malden MA and Oxford UK, 2017, pp. 179–194.

ed. Tirosh-Samuelson, Hava, *Judaism and Ecology: Created World and Revealed Word*, Cambridge, MA, 2002.

Treanor, Brian, *Emplotting Virtue: A Narrative Approach to Environmental Virtue Ethics*, Albany, 2014.

Tronto, Joan C., *Moral Boundaries: A Political Argument for the Ethics of Care*, New York, 1993.

Tronto, Joan C., *Caring Democracy: Markets, Equality and Justice*, New York, 2013.

Troster, Lawrence, »Tikkun Olam and Environmental Restoration: A Jewish Eco-Theology of Redemption«, *Jewish Education News*, 2008, http://www.caje.org.

Tucker, Richard, *Insatiable Appetite: The United States and the Ecological Degradation of the Tropical World*, Berkeley, 2000.

van Wensveen, Louke, *Dirty Virtues: The Emergence of Ecological Virtue Ethics*. Amherst, 2000.

van Wensveen, Louke, »The Emergence of Ecological Virtue Language«, in: *Environmental Virtue Ethics*, ed. Ronald Sandler and Philip Cafaro, Lanham, 2005, pp. 15–30.

ed. Warren, Karen J., *Ecofeminism: Women, Culture, Nature*, Bloomington, 1994.

Warren, Karen J., *Ecological Feminist Philosophies*. Bloomington, 1996.

Waskow, Arthur, »What Is Eco-Kosher«? in: *This Sacred Earth: Religion, Nature, Environment*, ed. Roger S. Gottlieb, New York and London, 1996, 297–300.

Werner, Micha H., »The Immediacy of Encounter and the Dangers of Dichotomy: Buber, Levinas, and Jonas on Responsibility«, in: *The Legacy of Hans Jonas: Judaism and the Phenomenon of Life*, ed. Hava Tirosh-Samuelson and Christian Wiese, Leiden, 2008, pp. 203–230.

Weiss, Raymond L., *Maimonides' Ethics: The Encounter of Philosophic and Religious Morality*, Chicago, 1991.

White, Lynn Jr., »The Historical Roots of Our Ecologic Crisis«, *Science*, Vol. 155, 1967, pp. 1203–1207.

Whyte, Kyle Powys and Chris J. Cuomo, »Ethics of Caring in Environmental Ethics: Indigenous and Feminist Philosophies«, *The Oxford Handbook of Environmental Ethics*, ed. Stephen M. Gardiner and Allen Thompson, Oxford UK, 2017, pp. 234–247.

Williston, Byron, *The Anthropocene Project: Virtue in the Age of Climate Change*, Oxford, 2015.

Winkler, Gershon, *The Place Where You Are Standing Is Holy: A Jewish Theology on Human Relationship*, Hoboken, 1998.

Winkler, Gershon, *Magic of the Ordinary: Recovering the Shamanic in Judaism*, Berkeley, CA, 2003.
Wyschogrod, Michael, »Judaism and the Sanctification of Nature«, *The Melton Journal*, Vol. 24 1991, pp. 5–7 (Reprinted in *Judaism and Environmental Ethics: A Reader*, ed. Martin D. Yaffe, Lanham, 2001, pp. 289–296).

Notes

[1] Zelig Golden is the co-founder and leader of Wilderness Torah, a Jewish environmental organization located in Berkeley, California, that seeks to actualise the vision of Zalman Schachter-Shalomi. This is just one example of contemporary Jewish attempts to reenchant nature.
[2] Tronto's understanding of political theory was shaped by her teacher Sheldon Wolin, a Jewish political theorist.
[3] Christine Cuomo (1998) differentiates between ecofeminism and ecological feminism but in this essay the terms are used interchangeably.
[4] The same holds true for environmental ethics: the more it shifted from debates about the ontological conditions for environmental ethics to the environmental needs and concerns of real communities, the more environmental ethics became attentive to religion. Within environmental ethics those who advocated environmental pragmatism (Light and Katz 1996; Minteer 2011) made it possible for the secular discourse of environmental ethics to pay attention to religion because communities use religious narratives. The significance of narratives for inspiring, motivating, and instructing is also recognised by proponents of environmental virtue ethics (Treanor 2014).
[5] In *Dirty Virtues* van Wensveen (2000) listed 189 environmental virtues and 174 environmental vices! The list presented here is considerably shorter.
[6] The Earth Charter (2000), which has been endorsed by thousands of organizations, hundreds of city governments, several national governments and UNESCO, states as its second principle: »Care for the community of life with understanding, compassion, and love.« Clearly the ethics of care and responsibility has already been shown to cross religious and secular boundaries.

Christliche Umweltspiritualität als Antwort auf die Umweltkrise

Die negativen Nebenwirkungen des technisch-wirtschaftlichen Fortschritts lehren uns je länger desto mehr das Fürchten. Zudem überschatten sie zunehmend die positiven Seiten der naturwissenschaftlichen, technischen und organisatorischen Erfindungen, ohne die eine Weltbevölkerung von gegenwärtig 7,7 Milliarden nicht überleben könnte. Dramatische Umweltprobleme, vor allem der Klimawandel, stellen diese modernen, technisch-naturwissenschaftlichen Entwicklungen, ja mehr die Moderne selbst und das ihr entsprechende menschliche Selbstverständnis fundamental in Frage. Um den gewaltigen Dynamiken und entfesselten Kräften globaler Weltentwicklung gerecht zu werden, bedarf es – dies ist offenkundig – eines grundlegenden Umdenkens, einer »mutigen, kulturellen Revolution«, wie Papst Franziskus in seinem Umweltschreiben formulierte.[1] Da die westliche Zivilisation mit ihren Erfolgen im Kampf gegen Hunger, Elend und Krankheiten aber auch ihren Schattenseiten inzwischen global geworden ist, bedarf es auch der globalen Bemühungen und Reflexion aller Akteure, einschließlich der Religionsgemeinschaften, um Ansätze aufzuzeigen, wie die ökologische Krise überwunden oder wenigstens abgemildert werden kann. Dies setzt vor allem voraus, dass möglichst viele Menschen weltweit den Weg zu einem neuen, nicht ausschließlich utilitaristischen Naturverständnis und der Achtung der Natur auch um ihrer selbst willen finden. Da kein gangbarer Weg hinter die modernen Entwicklungen zurückführt, es sei denn um den Preis gewaltiger humanitärer Katastrophen, stellt sich die dringliche Frage, welchen Beitrag Ethik und Spiritualität der Religionen zu einer Sicht der Natur leisten können, die der Realität begrenzter Ressourcen entspricht.

Ein Klassiker der Umweltethik, das »Prinzip Verantwortung«, hatte genau diese Frage gestellt. In ihm schildert Hans Jonas die durch die moderne Industrialisierung entstandene historisch neue Situation und bringt das Dilemma höchst präzise auf den Punkt:

Denn eben dieselbe Bewegung, die uns in den Besitz jener Kräfte gesetzt hat, deren Gebrauch jetzt durch Normen geregelt werden muß – die Bewegung des modernen Wissens in Gestalt der Naturwissenschaft – hat durch eine zwangsläufige Komplementarität die Grundlagen fortgespült, von denen Normen abgeleitet werden könnten, und hat die bloße Idee von Norm als solcher zerstört […]. Erst wurde durch dieses Wissen die Natur in Hinsicht auf Wert ›neutralisiert‹, dann auch der Mensch. Nun zittern wir in der Nacktheit eines Nihilismus, in der größte Macht sich mit größter Leere paart, größtes Können mit geringstem Wissen davon, wozu. Es ist die Frage, ob wir ohne die *Wiederherstellung der Kategorie des Heiligen* [Hervorhebung IG], die am gründlichsten durch die wissenschaftliche Aufklärung zerstört wurde, eine Ethik haben können, die die extremen Kräfte zügeln kann, die wir heute besitzen und dauernd hinzuerwerben und auszuüben beinahe gezwungen sind (Jonas 1980: 57).

Das Christentum als jene Religion, zu der sich mehr als ein Drittel der Menschheit bekennt, hat bereits aufgrund der Zahl seiner Gläubigen eine besondere Verantwortung. Zudem bildet es einen prägenden Faktor der technisch-naturwissenschaftlichen Zivilisation des Westens, die inzwischen global geworden ist. Wiewohl die vielfach vertretene Ansicht, dass die christlich-jüdische Tradition an der Umweltkrise die Schuld trage, nicht haltbar ist und eine unbegründete Spekulation darstellt (s. unten), stellt sich dennoch die Frage, wieso gerade in der westlichen Zivilisation die Problematik in dieser Schärfe entstehen konnte und welchen Beitrag der christliche Glaube für eine Umweltethik und -spiritualität leisten kann, die das menschliche Bewusstsein an ökologisch verträglichen Zielen ausrichtet. Eben dieser Frage soll im Folgenden in drei Punkten nachgegangen werden: Welche Bedeutung kommt dem Schöpfungsglauben für einen verantwortlichen Umgang mit der Umwelt zu? Welches sind die Grundlagen einer christlichen Anthropologie und Verantwortungsethik, die für eine Begründung einer Umweltethik wesentlich sind? Sowie abschließend: Welche Grundeinstellungen (Tugenden) aus christlicher Inspiration können dazu beitragen, ein neues Verhältnis zur Natur und Umwelt zu gewinnen, wobei zu zeigen sein wird, dass sich hier wesentliche Überlappungen mit anderen religiösen Traditionen aufweisen lassen.

1 Die Welt als Schöpfung: Zur Bedeutung biblischer Kosmologie für ein (christliches) Weltverständnis und seine Ethik

Der für die monotheistischen Religionen (vor allem Judentum, Christentum, Islam) charakteristische Glaube an die Geschaffenheit der Welt und des Menschen durch einen Gott besagt, dass dieser Gott, der Ewige, in Freiheit eine Welt außerhalb seiner selbst hervorbringen wollte und hervorbrachte *(creatio ex nihilo)*.[2] Dieser Schöpfungsakt wird dabei niemals als etwas Vergangenes und damit Abgeschlossenes verstanden, sondern ist im Sinne eines kontinuierlichen Wirkens Gottes in der Welt und Schöpfung zu begreifen, das sich durch die Zeit hin durchhält *(creatio continua)* und an dem der Mensch als Mitschöpfer mitzuwirken befähigt und beauftragt ist.

Für Max Weber, der in seiner »Wirtschaftsethik der Weltreligionen« die Rückfrage nach den kulturellen wie ethischen Wurzeln westlicher Zivilisation stellte, ist es erst der jüdische Monotheismus, der aufgrund seines Schöpfungsverständnisses einen rationalen und moralisch verantwortbaren Zugang zur Welt ermöglicht. Durch die kulturgeschichtlich schrittweise Zurückdrängung eines mythischen Weltbilds mit seinen durch vielfältige Opfer zu befriedigenden, unsichtbaren Zwischenwesen konnten – so Weber – jene Freiheitsräume entstehen, die die Voraussetzung für »eine religiöse Ethik des innerweltlichen Handeln« bildeten und die »noch der heutigen europäischen und vorderasiatischen Ethik zugrunde« liegen (Weber 1988: 6). Aufgrund der Entmachtung der dem Menschen wohlwollend oder feindlich gesinnten, in jedem Fall eigene Ansprüche anmeldenden Geistwesen, wie Feen, Dämonen, aber auch unterschiedlichste Götter, konnten die Gläubigen das Vertrauen gewinnen, dass die von *einem einzigen* Gott geschaffene und ihm untergeordnete Welt nach verlässlichen Gesetzen funktioniert. Diese grundsätzliche Kohärenz der Realität bildet die Voraussetzung für die in allen monotheistischen Religionen vorfindbare Überzeugung, dass der (gläubige) Mensch nur diesem einen Gott gegenüber verantwortlich ist, was ethische Rationalität zuallererst ermöglicht.[3] Die Entgötterung der Welt im jüdischen und später christlichen Glauben und Denken (ebenso wie nochmals später im islamischen), verbunden mit dem Verbot der Magie, bildet daher die Grundlage für eine ethisch-rationale Lebensführung. Nur in einer grundsätzlich berechenbaren Welt ist es zudem möglich, universale Naturgesetze und ihre technische Anwendung zu erkennen, die zudem, weil eben die ganze Welt von Gott geschaffen ist, universale Geltung beanspruchen können. Wiewohl mit dieser Erkenntnis aus heutiger Sicht beachtliche Probleme einhergehen, gilt es dennoch, diese Grundlagen universaler Rationalität zuerst zu verstehen, um dann in einem zweiten Schritt die Frage zu stellen, was dies für die welthistorisch neue Situation weit verbreiteter Umweltzerstörung bedeutet.

Zwei biblische Narrative, die so genannten Schöpfungserzählungen, die sich beide im ersten Buch der Bibel, dem Buch Genesis (Entstehung der Welt) finden, explizieren diese Überzeugung von der Erschaffung der Welt durch den einen Gott. Der erste, längere Text stammt aus dem 6. Jhdt. v. Chr., der zweite – ihm nachgeordnete – aus dem 10. Jhdt. v. Chr. Diese beiden, inhaltlich höchst unterschiedlichen Erzählungen stehen ohne Harmonisierung hintereinander und interpretieren sich wechselseitig. Vom Genus her haben beide ätiologischen, d. h. erklärenden, Charakter. Ihr Ziel ist es, die göttlichen Intentionen im Hinblick auf den Kosmos und den Menschen aufzuzeigen. Sie sind so universal auf die gesamte Menschheit bezogen. Sie sind zugleich den historisch verortbaren Geschichten, die mit der Abrahamserzählung in Genesis 12 beginnen und die Yahwes erlösendes Handeln am Volk Israel behandeln, vorangestellt. Die Schöpfungserzählungen legen demnach den universalen und kosmologischen Rahmen dieser als Heilsgeschichte begriffenen, folgenden Texten fest. Sie tun dies, indem sie die Grundgegebenheiten des vorfindbaren Kosmos sowie der *conditio humana* in doppelter Weise theologisch reflektieren: Den ersten Teil bildet eine Beschreibung der Welt, wie sie nach göttlicher Intention sein *sollte*, den zweiten die real erfahrbare, deformierte Weltwirklichkeit. Die Bruchstelle zwischen diesen beiden Teilen markiert die, gleichfalls ätiologische, Sündenfallerzählung (Genesis 3), auf die der Mord an Kain (Genesis 4) sowie die Sintfluterzählung (Genesis 5/6–9) folgt. Sie endet mit der Zusicherung, dass die Schöpfung trotz der moralischen Verderbnis der Menschen Bestand haben wird. Damit ist auch klar, dass es sich bei den Schöpfungserzählungen nicht um einen Evolutionsbericht handelt. Sie enthalten vielmehr eine Kosmologie und eng damit verbunden eine Anthropologie, deren Basis die Erschaffung des Kosmos durch Gott in sechs Tagen mit dem siebenten Tag als Ruhetag (Sabbat) bildet. Bereits hier wird offenkundig, dass es im Text nicht um Zeitangaben geht (vgl. Zenger 1987). Vielmehr wird das Narrativ über die Erschaffung der Natur einschließlich des Menschen mit der aus der hebräischen Kultur stammenden, gleichfalls biblisch vermittelten Wochenperiodisierung verschränkt. Die Schöpfung durch Yahwe wird zudem analog zur menschlichen Arbeit gedacht, der damit auch eine besondere Bedeutung zugesprochen wird. Ihren Gipfel- und Höhepunkt bildet der Sabbat als Ruhetag, durch den das Schaffen Gottes wie des Menschen geistig überhöht und damit bei aller Wichtigkeit und Würde der Arbeit relativiert wird.[4] Auch hier zeigt sich – allgemein gesprochen –, dass Natur und Kultur unauflösbar miteinander verschränkt sind. Die Art und Weise wie Menschen ihre natürliche Umwelt begreifen, ist demnach immer auch kulturell vermittelt und notwendig in eine das gesamte Denken und Handeln umfassend prägende Weltsicht eingebettet.

Dazu einige weitere Anmerkungen im Hinblick auf die Umweltthematik. Die sogenannte priesterschriftliche Schöpfungserzählung aus dem 6. Jhdt., die den Anfang der Bibel bildet, beginnt mit den Worten: »Am Anfang schuf Gott Himmel und Erde. Die Erde aber war wüst und leer [hebräisch *tohu wabohu,* ›dunkle weglose Wüste‹] und Finsternis war über der Urflut […] und Gottes Geist [hebräisch *ruah,* weitere Bedeutungen: »Leben«, »Atem«] schwebte über den Wassern.« (Genesis 1,1) Die bereits vorhandene ungeformte Materie wird so erst durch die göttliche Schöpfungstat zum gestalteten Kosmos. Dieser Prozess beginnt mit der Erschaffung von Licht und Finsternis (erster Tag), gefolgt von der Schöpfung von Himmel, Erde und Meer, die – wie allgemein im Alten Orient – als drei örtliche Sphären vorgestellt werden (zweiter Tag), den Pflanzen (dritter Tag) sowie Sonne, Mond und Sternen (vierter Tag). Letztere sollen den Wechsel der Zeiten und Feste anzeigen. Hier zeigt sich wiederum die innere Verschränkung von Kultur und Natur im Schöpfungstext. Die Erschaffung der Wassertiere und Vögel (fünfter Tag) wie auch der übrigen Tiere, der Land- und Wildtiere (sechster Tag), wird jeweils mit einem so genannten Vermehrungssegen verbunden, der nicht zuletzt aufzeigt, dass ihr Bestand mehr gefährdet ist als jener der anderen Kreaturen. Dies gilt auch für den Menschen, der gemeinsam mit einem Großteil der Tierarten am sechsten und letzten Schöpfungstag erschaffen wird. Dies erlaubt eine erste für die gegenwärtige Umweltthematik äußerst relevante Einsicht. Der Mensch ist nach biblischem Verständnis zuerst und vor allem in das Gesamt der physischen Schöpfung, der Natur, eingebunden. Seine Sonderstellung ergibt sich freilich aus dem abschließenden Zusatz: »Als Abbild Gottes *(säläm)* schuf er ihn. Als Mann und Frau schuf er sie« (Genesis 1,27). Damit ist bereits die nur für den Menschen typische Zwischenstellung aufgezeigt: Er/sie gehört zu den höheren Tieren.[5] Zugleich aber kommt ihm/ihr als Abbild *(sälām)* des Gottes eine Sonderstellung in der Natur und im Kosmos zu. Diese wird mit dem Wort *sälām*, das »Statue«, »Bildnis« bedeutet, charakterisiert. Es verweist damit auf die allgemein im Alten Orient übliche Praxis, die Statue eines Herrschers (oder Gottes) auf dem Stadtplatz (oder im Tempel) aufzustellen, um diesen Herrscher oder Gott dort präsent zu setzen. Der Mensch repräsentiert, so die Grundaussage des Textes, die Gottheit respektive den Herrscher (wobei beides oft in eins fällt). Diese Repräsentanz ist dabei nicht zuerst und vor allem als Seinsaussage zu verstehen, sondern ihr kommt primär ethische Bedeutung zu. Das Leben des Adam (*adamah*, »Menschlein«) soll und muss von anderen Menschen respektiert werden, da er/sie Abbild Gottes sind. Die negative Parallelerzählung dazu bildet die Ermordung Abels durch Kain nach dem Sündenfall, also in einer deformierten, permanent von Mord bedrohten Welt. In den jüdischen wie christlichen Traditionen bildet diese Grundeinsicht das Fundament der gesamten Anthropologie und ist das Grundaxiom der Ethik (Janowski und Liess 2009; Frevel 2010; Konradt und Schläpfer 2014). Im Menschen als Geschöpf verbinden sich so Naturhaftigkeit, d.h. Lebendigkeit, Verwundbarkeit, Endlichkeit und letztlich Sterblichkeit, mit jener spezifischen Transzendenz, die als Gottesebenbildlichkeit bezeichnet wird und die es aufgrund des sogenannten noachitischen Grundgebots verbietet, sein Blut zu vergießen (Genesis 9,5). Aus der Gottesebenbildlichkeit werden in der Folge spezifisch menschliche Fähigkeiten abgeleitet, wie Verstand, Wille und individuelle Kreativität (Thomas von Aquin 1985). Diese anthropologische Deutung bleibt jedoch eng verbunden mit der Sicht des Menschen als eines Gott repräsentierenden Gegenüber, jenes Anderen, dessen spezifische Würde aufgrund seiner/ihrer Gottähnlichkeit zu respektieren die Grundlage der Ethik bildet (vgl. dazu Levinas 1989).

Sie bildet zusammen mit der Vermehrungsverheißung und dem so genannten Herrschaftsauftrag eine inhaltliche Einheit, weshalb die Aussagen textlich miteinander verschränkt sind: »Gott segnete sie und Gott sprach zu ihnen: Seid fruchtbar, und vermehrt euch, bevölkert die Erde, unterwerft sie euch und herrscht über die Fische des Meeres, über die Vögel des Himmels und über alle Tiere, die sich auf dem Land regen.« (Genesis 1,28, Vorverweis in Genesis 1,26, weiters Genesis 9,1 nach der Sintflut).[6]

Da dieser Vers in den letzten Jahrzehnten vielfach direkt für die Umweltkatastrophen unserer Zeit und eine westliche Mentalität des Beherrschens verantwortlich gemacht wurde, ist hier in gebotener Kürze auf die diesbezügliche Kritik einzugehen, die sich vor allem gegen das Christentum richtet. Es gibt mehrere Gründe, die diese These trotz ihrer Popularität als nicht haltbar erscheinen lassen. Da ist zum ersten der Text selbst. Die hebräische Wortwurzel *raddah* bedeutet »führen«, »auf Weideland geleiten« und geht damit von einem sich am Beruf des Hirten orientierenden Herrschaftsverständnis aus, wie es für den Alten Orient typisch war.[7] Im Hintergrund steht demnach nicht das Bild eines orientalischen Despoten, sondern eines fürsorglichen Königs. Diese Deutung gewinnt dadurch an Gewicht, dass die zweite, zeitlich frühere Schöpfungserzählung (Genesis 2,4b-24) davon spricht, dass Gott den Mann und dann die Frau aus Lehm formte, ihnen den Lebensodem einblies und ihnen die Welt als einen zu bebauenden und zu pflegenden Garten, den Garten Eden, zur Kultivierung übertrug. Das Herrschaftsmotiv von Genesis 1, 28 wird hier textimmanent im Sinne der Fürsorge für das Land interpretiert. Zwar hat der Mensch ein Vorrecht, das darin besteht, den anderen Geschöpfen ihre Namen zu geben, was auf die Sonderstellung des Menschen als Vernunftwesen verweist (Genesis 2,19), aber die Grundintention des Textes ist es, die Besonderheit des Menschen im Sinne seiner Verantwortung für die Schöpfung zu interpretieren. Die islamische Vorstellung, dies sei hier nur erwähnt, ist im Übrigen weitgehend analog. Gott der Schöpfer aller Dinge setzt den Menschen als *khalifa* (Statthalter) über seine Schöpfung ein, für die er/sie eine Art Treuhandschaft übernimmt. Auch für das koranische Denken liegt die Umweltproblematik außerhalb des eigenen Denkhorizonts und wird Herrschaft nicht willkürlich absolutistisch sondern fürsorglich verstanden.[8] Den allen Monotheismen gemeinsame und für sie charakteristische Vorstellung ist demnach trotz aller Unterschiede im Einzelnen, dass der Schöpfergott den Menschen geschaffen und zugleich beauftragt hat, gerecht über seine Schöpfung zu herrschen. Er soll so die gerechte Herrschaft Gottes, der für seine Geschöpfe Sorge trägt, nachahmen und weiterführen. Die ökologische Theologie spricht daher zu Recht von einer menschlichen Treuhandschaft *(stewardship)* über die Natur, die in direktem Gegensatz steht zu jeder Form der Willkürherrschaft und Ausbeutung.

Hinzu kommt ein Zweites: Die Herleitung der naturwissenschaftlich-technischen Erfindungen der Moderne aus einem biblischen Halbsatz ist nicht nur bibelinterpretatorisch unzulässig, sie ist auch schlechterdings anachronistisch. Denn sie übersieht, dass über Jahrtausende der Mensch gegenüber der Natur in einer höchst prekären Lage war. Als Mängelwesen (Arnold Gehlen) hatte er in jener agrarischen Umwelt, aus der die biblischen Texte stammen, Jahr für Jahr um sein Überleben zu kämpfen. Ernteausfälle, Ungeziefer, Dürre u. Ä. m. führten immer wieder zu Hungersnöten und bedrohten das menschliche Leben, ja jenes der Spezies (so auch Genesis 6–9). Jene weltgeschichtliche Wende, durch die der Mensch nun seine eigene Existenz und jene der anderen Lebewesen bedroht, war für die biblischen Autoren des 1. Jahrtausends v. Chr. – so könnte man sagen – ebenso wenig vorstellbar wie die Konstruktion von Flugzeugen, Mikrochips und Atombomben.

Eine Entzauberung der Welt, der nichts Innerweltliches mehr heilig ist (Hardmeter und Ott 2016: 183), lässt sich demnach in keiner Weise aus dem Text der biblischen Schöpfungserzählung herauslesen. Sie stellt vielmehr das Resultat einer langen Reihe geschichtlicher Transformationsprozesse dar, die keineswegs linear und zwangsläufig erfolgten. Erst in der letzten Etappe dieser komplexen Entwicklungsgeschichte von mehreren Jahrtausenden, die offenkundig an unterschiedlichen Orten zu unterschiedlichen Ergebnissen führten, wurde die Sonderstellung des Menschen gegenüber der Natur absolut gesetzt und eine strikte Trennlinie zwischen ihm und dem Rest der Natur eingezogen. Im Gegensatz zum biblischen Weltbild reduzierte diese cartesianische Subjekt-Objekt-Dichotomie die doppelte Ausrichtung menschlicher Existenz auf *eine einzige* Dimension, jene eines der Natur gegenüberstehenden (transzendenten) Bewusstseins, das in der menschlichen Ratio gründet. Die offenkundig naturhaft-biologische Dimension des Menschen wird damit ausgeblendet. Er wird nicht mehr als Teil der Natur gesehen sondern als ein rein geistiges Subjekt, das aufgrund seiner Vernunft über dieser steht. Aus seiner natürlichen Verankerung gänzlich herausgelöst begreift er/sie sich nun als Gegenpol zur nunmehr rein physisch verstandenen Materie, die ihres Schöpfungscharakters entkleidet ist. Damit geht freilich auch ihr Eigenwert verloren, der nach biblischem Verständnis der nicht-menschlichen Schöpfung in allen ihren Formen zukommt. Sie wird zum reinen Objekt, über das der Mensch zum eigenen Nutzen absolut verfügen kann. Das anthropologische Leitbild der Moderne versteht den Menschen daher als Besitzer, Arbeiter und Forscher, der die vorhandene Natur willkürlich beherrschen kann und soll (Barruzi

1993: 56). Die Arbeit wird zur »Erzeugerin und Erlöserin« eines neuen Menschen (Leser 1980: 67), der sich so von seiner eigenen biologischen Natur zu emanzipieren sucht.[9] Erst hier wird das Verhältnis zur Natur als ein Herrschaftsverhältnis gefasst, das die bereits aufgrund der gemeinsamen biologischen Konstitution vorhandene innere Verbundenheit des Menschen mit der übrigen Natur zerschneidet. Anders in den biblischen Schöpfungserzählungen, für die die Interdependenz alles Lebendigen aufgrund seines Geschaffenseins gegeben ist und die die Sonderstellung des Menschen immer im Zusammenhang mit seiner mit allen anderen Lebewesen gemeinsamen biologischen Existenz denken. Dies zeigen im Übrigen viele Gebetstexte, die das Lob des Schöpfers durch die gesamte Natur zum Inhalt haben und dessen universale kosmische Dynamik. Zur Illustration seien einige Verse zitiert:

> Die Himmel rühmen die Herrlichkeit Gottes, vom Werk seiner Hände kündet das Firmament. Ein Tag sagt es dem anderen, eine Nacht tut es der andern kund, ohne Worte und ohne Reden, unhörbar bleibt ihre Stimme, doch ihre Botschaft geht in die ganze Welt hinaus, ihre Kunde bis zu den Enden der Erde (Psalm 19, 2–7).

Gerade die als Gebete für Juden wie Christen zentralen Psalmen zeigen somit, dass biblisch gesehen, die Natur alles andere als unbelebte Materie ist. Ihre Poesie könnte dazu beitragen, den Eigenwert aller Geschöpfe wieder neu zu entdecken.[10] Gleiches gilt für die Schriften christlicher Mystiker wie des Franz von Assisi (1181/1182–1226) oder von Angelus Silesius (1624–1677).[11]

2 Der anthropologische Zugang: Der Mensch als Mitschöpfer und Bewahrer einer Schöpfung in Evolution

Ebenso wie der Subjekt-Objekt-Dualismus ist für die Moderne zudem ein ihr bereits vom Begriff her eingestiftetes, geschichtsphilosophisches Fortschrittsmodell charakteristisch. Es ist einerseits deterministisch, anderseits programmatisch und zielt auf eine permanente Selbstüberbietung. Knapp gesagt: Das Neue ist immer und notwendig das Bessere. Diese Hypothese einer linearen Entwicklung der Weltgeschichte findet ihren Ausdruck paradigmatisch im Dreistadienmodell des Philosophen und Begründers der Soziologie, Auguste Comte (1798–1857). Die Geschichte bewegt sich demnach unaufhaltsam vom religiös-christlichen Obskurantismus über eine philosophische Weltsicht hin auf ein positives, aufgeklärtes Zeitalter. Es ist eben diese Grundannahme eines linearen Fortschritts, die aus religiöser und insbesondere christlicher Sicht nicht überzeugt und heute nicht zuletzt aufgrund der Umweltprobleme in eine fundamentale Krise geraten ist.

Ihre Schwäche bestand von Anfang an darin, dass sie ihre »barbarische Rückseite« (Habermas 1994: 52),[12] und damit auch die negativen Folgen naturwissenschaftlich-technischer Innovationen, auszublenden beinahe gezwungen ist. Diese inhärente Einseitigkeit führt je länger desto mehr zu einer Haltung, die jener eines Spielers in einem Casino gleicht, der immer höhere Risiken auf sich zu nehmen bereit ist, um das Spiel doch noch zu gewinnen. Der Soziologe Ulrich Beck hat in eben diesem Sinn von der »Risikogesellschaft« als Charakteristikum der Spätmoderne gesprochen (Beck 2015). Der Fortschrittsglaube der Moderne wird dabei Schritt für Schritt von einer Hoffnungsvision konstanter Humanisierung zur Vorstellung eines »stählernen Gehäuses« (Max Weber), die das eigene Weltbild unterhöhlt.

Diese der Moderne von Anfang an inhärente Problematik wurde bereits früh erkannt. Dies zeigen Gedichte wie der Zauberlehrling von Johann W. Goethe sowie die romantischen Gegenentwürfe, die zu Beginn des 19. Jahrhunderts an Bedeutung gewinnen. Sie setzen dem rationalistischen und fortschrittsorientierten Narrativ der Moderne eine »Bewegung der Ganzheitlichkeit« entgegen, rufen zu einer »Rückkehr zur Natur« auf und/oder idealisieren frühere Kulturformen, sei es jene der Antike oder des Mittelalters, als der Moderne überlegen. Die ursprüngliche Einheit mit einer idyllisch überhöhten Natur soll durch die »Wiederverzauberung« der Welt möglich werden. Mit dieser Idee verbindet sich bereits in der Romantik eine Kritik des Monotheismus und hier wiederum des Christentums, das für eben diese »Entzauberung« verantwortlich gemacht wird.[13] Der heutige Neoromantismus und die ihn begleitenden esoterischen Bewegungen sind Nachfahren dieser romantischen Weltsicht, die die Fortschrittshypothese durch eine nostalgisch unterfütterte Verfallshypothese ersetzte. Diese steht freilich vor der Frage: Kann der Glaube an eine von Geistern und Göttern bevölkerte Welt heute einen echten Beitrag zur Lösung der ökologischen Frage leisten? Oder handelt es sich nicht vielmehr um einen Eskapismus, der zur Flucht aus der Verantwortung verführt? Und was bedeutet dies angesichts einer wachsenden Weltbevölkerung, für die der Verfall der technischen Zivilisation humanitäre Katastrophen unvorstellbaren Ausmaßes mit sich bringen würde? Es erscheint damit offenkundig, dass es in der gegenwärtigen ökologischen Situation ein Höchstmaß an menschlicher Besonnenheit und kreativer Verantwortungsbereitschaft braucht, um mit den Folgen moderner Entwicklungen in einer Weise umzugehen, die die bereits vorhandenen Schäden minimiert und ihnen nicht weitere hinzufügt. Ein nostalgischer, auf eine so nie gegebene ursprüngliche Einheit zielender Naturmystizismus lenkt jedoch von eben dieser verantwortungsorientierten Sicht der Welt und einem ihr entsprechenden Tun ab.

Einen zweiten neben einer Einheitsromantik höchst einflussreichen Versuch, die Subjekt-Objekt-Dichotomie der Moderne zu überwinden, stellt ein biologistischer Evolutionismus dar. Wiewohl die Evolutionisforschung verschiedene Ausprägungen entwickelt hat (vgl. Sarasin 2010; Langthaler und Weber 2013), basieren sie doch durchgängig auf der Hypothese, dass der Mensch ein Säugetier unter anderen in der Kette einer Evolution darstellt, die durch Kampf und Selektion den *survival of the fittest* deterministisch herbeiführt. Nicht der Mensch ist es demnach, der der historischen Entwicklung ihre Richtung vorgibt, sondern ein sich selbst determinierender evolutionärer Fortschrittsprozess, dessen Objekt er ist. Ein derartiges Weltbild lässt offenkundig kaum Raum für menschliche Freiheit und damit Verantwortung. Es erscheint daher weder mit einem christlichen noch einem immanenten Humanismus als vereinbar, die beide von der Grundannahme menschlicher Freiheit nicht absehen können, wollen sie nicht jene Transzendenzvorstellung aufgegeben, die den Menschen – christlich – als Bild Gottes und philosophisch-ethisch als mit besonderer Würde begabt versteht. Die Attraktivität eines derartigen Evolutionismus liegt m. E. darin, dass er einen scheinbaren Ausweg aus der cartesianischen, einseitig auf die Vernunft fixierten Weltsicht aufzeigt, sowie die für die Moderne zentrale Fortschrittsidee dadurch rettet, dass er sie evolutionistisch deutet. Sein gegenwärtiger Boom, für den Autoren wie Steven Pinker (2016, 2018) und Yuval Noah Hariri (2017, 2018) stehen, scheint daraus erklärlich.

Der Paläontologe, Christ und Jesuit Pierre Teilhard de Chardin hat bereits in der ersten Hälfte des 20. Jhdt. eine alternative Evolutionsthese vorgelegt. Sie kann hier nicht näher ausgeführt werden, könnte jedoch einen Weg der Neubewertung von Fortschritt aufzeigen und soll daher erwähnt werden. Die gegenwärtige Phase der Anthropogenese (heute des Anthropozäns) fordert nach Teilhard de Chardin ein menschliches Handeln heraus, das in freier Entscheidung die humanen Potentiale der Schöpfung weiterentwickelt. Der »Mensch im Kosmos«, so der Titel seines Hauptwerks, und seine weitere Evolution hängen von der Wahrnehmung eben dieser ethischen Verantwortung durch die gegenwärtige Generation ab (de Chardin 1959). Fortschritt stellt demnach keine automatische, evolutionär bedingte oder auf natürlicher Selektion basierende Notwendigkeit dar. Er muss vielmehr durch das Handeln jedes einzelnen Menschen unter Einsatz aller seiner geistigen Energien und schöpferischen Kräfte weitergeführt werden. Sein Letztkriterium sind die Personwerdung jedes einzelnen Menschen, Mann und Frau, sowie das allgemeine Wohl der Welt in der Realisierung des Humanen (de Chardin 1969: 132). »Fortschritt heißt menschlicher sein oder er bedeutet nichts«[14] – so die Grundaussage (de Chardin zitiert in Broch 1977: 322). Ein derartiger Weltfortschritt hat für den Christen Teilhard de Chardin sein letztes Maß in Christus, der den Menschen in seiner vollendeten Gestalt darstellt und Grund wie Ziel der gesamten Menschheit und des Universums ist. Fortschritt wird hier von einer geschichtsnotwendigen Größe zum Resultat menschlichen Handelns umgedeutet, das sich für die Humanisierung der Welt in allen ihren Dimensionen verantwortlich weiß. Es ist die Aufgabe der Religion, diese Art des Fortschrittes und damit die Quellen des *élan vital*, der Lebens- und Handlungsfreude, zu fördern (de Chardin 1963: 245; vgl. auch Broch 1977: 226–234). Ein derartiges, nicht deterministisches und nicht auf dem Willen zur Macht basierendes Fortschrittsmodell lässt sich mit dem Christentum und seiner Ethik verbinden. Positive wie negative Entwicklungen sind demnach keineswegs zwingend, wiewohl es Voraussetzungen für ihre Realisierung gibt. Diese Ernstnahme menschlicher Freiheit führt eben dazu, dass massive Rückschritte und apokalyptische Entwicklungen ebenso möglich sind wie Fortschritte. In diesem Sinn schreibt Papst Franziskus in seinem Antrittsschreiben *Evangelii gaudium:* »Wir stehen hier vor einer großen Verantwortung, weil einige gegenwärtige Situationen, falls sie keine guten Lösungen finden, Prozesse einer Entmenschlichung auslösen können, die dann nur schwer rückgängig zu machen sind« (Papst Franziskus 2013).

Die Frage, die sich im Anschluss daran stellt, ist freilich: Wie lassen sich nach Jahrhunderten im menschlichen Bewusstsein fest verankerte Mentalitäten eines rein instrumentellen Umgangs mit der Natur nachhaltig verändern? Wie lässt sich die etwas rezentere, aber nun weit fortgeschrittene Kolonialisierung aller Lebensbereiche durch Naturwissenschaften, Technik und Ökonomie aufhalten? Wie kann eine Weltsicht gefördert werden, in der sich der Mensch nicht als absoluter Herrscher über die Natur, sondern als Hüter der Erde von seiner Verantwortung her begreift? Anders gesagt, wie lässt sich jene cartesianische Spaltung rückgängig machen, die letztlich auch den Menschen als *res cogitans* reifiziert und entpersönlicht, ohne damit freilich eine naturwissenschaftlich-technische Entwicklung zu verunmöglichen?

Im abschließenden Abschnitt soll es darum gehen, einige jener christlichen und allgemein religiösen Haltungen herauszuarbeiten, die einem zerstörerischen Umgang mit der Natur und ihren Ressourcen entgegenwirken könnten. Eine derartige »Tugendethik« steht, so sei vorweg angemerkt, in keiner Weise im Gegensatz zu struktur- und rechtsethischen Überlegungen und sie politisch effektiv umsetzenden Maßnahmen.[15] Es bedarf heute mehr denn je der Institutionen, Gesetze und Regelungen zum Schutz der Umwelt auf globaler, nationaler und regionaler Ebene. Ohne sie können der Natur bereits zugefügte Schäden nicht reduziert und die ärmeren Bevölkerungen der Welt, die den Auswirkungen der Umweltzerstörung am stärksten ausliefert sind, nicht wirksam geschützt werden.[16] Dies gesagt, sollen im Folgenden individuelle Haltungen benannt werden, die als Grundlagen auch für strukturethische Fortschritte unverzichtbar erscheinen.[17]

3 Ökologisch relevante Tugenden aus christlich-interkonfessioneller und interreligiöser Perspektive

a. Humilitas als Anerkennung der Begrenztheit menschlicher wie naturaler Wirklichkeiten

Die Thematisierung von Grenzen, einschließlich der begrenzten Lebenszeit und der materiellen Ressourcen ist höchst überraschend kein Bestandteil moderner Theorien. Diese Schwach-und Leerstelle bedeutet letztlich einen Verlust des Bezugs zu einer menschlichen wie naturalen Wirklichkeit, die immer und überall durch Verwundbarkeit, Unvollkommenheit und letztlich Sterblichkeit charakterisiert ist, zu der beim Menschen aufgrund seiner Freiheit seine Fehlbarkeit (biblisch Sündhaftigkeit) kommt. Diese offenkundigen und unumstößlichen Tatsachen stellen jedoch einen notwendigen Bezugspunkt ethischer Rationalität dar. Ihre Anerkennung kann mit dem in der christlichen Ethik ursprünglich verankerten Begriff der *humilitas* bezeichnet werden.[18] Das Wort leitet sich ursprünglich von *humus,* »Erde«, ab. Der Mensch ist demnach als Teil der Natur und ihrer Vergänglichkeit auf die Erde in ihrer immer mitgegebenen Begrenztheit als natürliches Habitat angewiesen.

Die christliche Haltung der *humilitas* stellt den Gegensatz zur *hybris* (Hochmut) dar, die mit einer Überschreitung menschlicher Möglichkeiten und menschlichen Maßes einhergeht. Die durchaus moderne Vorstellung, dass der Mensch sich je neu selbst erschaffen kann und seiner Ingenuität keine Grenzen gesetzt sind, vermittelt eine Illusion absoluter Machbarkeit, die durch keinerlei reale Bedingungen gedeckt ist. Ohne die schöpferischen Fähigkeiten des Menschen gering zu achten, liegt ein Problem heutiger Vorstellungen von Naturbeherrschung eben darin, dass ihre grundsätzliche Intention ist, Grenzen jeweils neu zu überschreiten. Eine derartige Einstellung widerspricht einer selbstreflexiven *humilitas,* die realistisch die Begrenztheit der eigenen Möglichkeiten in der jeweiligen Situation in Denken und Handeln einbezieht und sich von der verführerischen Vorstellung der Grenzenlosigkeit nicht blenden lässt. Ein derartiger nüchterner Realismus, der eben als *humilitas* bezeichnet werden kann, wäre höchst notwendig, um ökologische Verbesserungen in allen Bereichen effektiv voranzutreiben. Eine Anerkennung der offenkundigen Begrenztheit menschlicher wie tierischer Existenz (wie der natürlichen Ressourcen) könnte zu einer neuen Sicht aller Geschöpfe und alles Geschaffenen, also der außermenschlichen Realität, führen, die deren intrinsischen Wert wieder klarer in den Blick bekommt.[19]

b. Maß und Verzicht als wirklichkeitsadäquate Haltungen

Die Haltung der *humilitas* erweist sich bei näherem Hinsehen eng verwoben mit jener des Maßes. Diese bildet bereits in der griechischen Philosophie neben der Besonnenheit die eigentliche menschliche Urtugend und wurde von dort in das frühe Christentum übernommen (Dihle 1957). Es gibt demnach ein allgemein menschliches Maß, das in unterschiedlichen Situationen und bei unterschiedlichen Menschen variieren kann, jedoch nie in eine grundsätzliche

Unbegrenztheit umschlagen kann. Dies eben wäre das Laster der *pleonexia* als Grundlage einer von Begehrlichkeit und Gier beherrschten Lebensweise. Es ist beachtenswert, dass das immer Mehr-Haben-Wollen, das hier wie in allen religiösen Ethiken als Laster gilt, im Wirtschaftsdenken der Moderne zur wesentlichen Triebkraft für den Fortschritt und damit zu einer Tugend avanciert. Seine sozial negative Seite wird dabei schlicht unterschlagen (Gabriel 2011). Alle Religionen sowie vormoderne philosophische Weltanschauungen gehen hingegen davon aus, dass die Maßlosigkeit des Einzelnen wie ganzer Gesellschaften das Zusammenleben und letztlich den Frieden gefährdet. Dies kann nur solange verborgen bleiben, als der technische und wirtschaftliche Fortschritt das rechte Maß im Gebrauch der Güter immer wieder kräftig nach oben verschiebt, wobei bereits früh Krisen und Revolutionen die Fragilität dieser wirtschaftlichen Dynamiken zeigen. Eine grundsätzliche Infragestellung erfährt diese Mentalität durch die Umweltkrise. Sich der Frage des Maßes ernsthaft zu stellen, erweist sich daher heute mehr denn dringlich. Dies umso mehr als offenkundig geworden ist, dass der Lebensstil der reicheren und industrialisierten Regionen aufgrund der Begrenztheit der natürlichen Ressourcen auf eine wachsende Weltbevölkerung nicht übertragbar ist.[20] Nur durch einen maßvolleren Umgang mit den natürlichen Ressourcen können längerfristig weltweite Verteilungskämpfe vermieden werden. Eine neue, positiv konnotierte Sicht des Maßes und der Selbstbeschränkung sind so aufgrund der Verpflichtung zu weltweiter Solidarität gefordert. Sie bilden jedoch auch die Grundlage innerer Freiheit und menschlicher Zufriedenheit. Die Weisheit aller Religionen lehrt, dass Einfachheit in der Lebensführung ein hohes Gut darstellt.[21] Das Ideal war und ist demnach eine Reduktion der materiellen Bedürfnisse und nicht ihre Ausweitung, die notgedrungen zur Entfremdung des Menschen von der Natur und seinen wesentlichen Lebenszielen führt. So schreibt der christliche Kirchenlehrer Basilius im 4. Jhdt. stellvertretend für viele Weisheitslehrer: »Arm ist, wer viele Bedürfnisse hat« (Basilius von Cäsarea 1925: 248). Diese Haltung liegt jedoch offenkundig quer zur Moderne. Gerade die ökologischen Probleme zeigen jedoch, dass Knappheiten in einer endlichen Welt nie überwunden werden können. Eine dies akzeptierende Grundhaltung erforderte eine Beschränkung, nicht eine Ausweitung der Bedürfnisse als Ausdruck von Solidarität wie Nachhaltigkeit.

Die Frage: Wie viel der Mensch braucht, um ein menschenwürdiges und gutes Leben zu führen, muss daher heute trotz wirtschaftlicher und technischer Dynamiken, die die Grenzen hinauszuschieben und den Konsum über jedes historisch bekannte Maß hinaus zu steigern erlaubten, aufgrund der evidenten Knappheit natürlicher Ressourcen neu gestellt werden.[22] Ein Weg dazu könnte eine Wiederbelebung der alten Fastenordnungen sein, wie sie sich in allen Religionen finden, um u.a. den Fleischverbrauch zu verringern.[23] Gelänge es, derartige Einsichten und religiös fundierte Praktiken zu gesellschaftlichen Trends zu verdichten, könnte ein Lebensstil gefördert werden, der der Begrenztheit der Güter der Welt Rechnung trägt und ihren Eigenwert wieder anerkennt.

c. *Dankbarkeit als Grundlage einer christlichen und relationalen Weltsicht*

Die moderne Vorstellung, wonach der Mensch als Individuum die Welt einschließlich seiner selbst erschafft, ist vielfach mit der Illusion verbunden, dass die eigene Existenz und der eigene Erfolg ausschließlich Folge der eigenen Leistung sind. Dies widerspricht jedoch offenkundig den sozialen und ökologischen Realitäten. Jeder Mensch lebt von Anfang bis Ende seines Lebens in einer Vielzahl sozialer Bezüge und ist auf eine von ihm selbst nicht hergestellte Natur wie Kultur, einschließlich ihrer Sprache angewiesen. Dankbarkeit bedeutet jene Grundhaltung, die diese der individuellen Existenz vorausliegenden sozialen, geschichtlichen und naturalen Interdependenzen wertschätzend anerkennt. Sie weiß um die Beschränktheit der je eigenen Leistung. Eine derartige Grundhaltung der Dankbarkeit ermöglicht es auch, die Natur in ihrer Eigenart und Schönheit zu begreifen, ohne sie zuerst und vor allem als Quelle eigener Bedürfnisbefriedigung zu vereinnahmen. Christlich verstanden wird sie so als Schöpfung verstanden und damit als Ausdruck von Gottes Wirken und Fürsorge. Dieses biblische und christliche Weltverständnis findet seinen poetischen Ausdruck gleichfalls in den Psalmen, wo es heißt: »Du sorgst für das Land und tränkst es, Du überschüttest es mit Reichtum […]. Der Bach Gottes ist reichlich gefüllt, Du schaffst ihnen Korn: so ordnest Du alles. Du tränkst die Furchen, ebnest die Schollen, machst sie weich durch Regen, segnest ihre Gewächse, Du krönst das Jahr mit deiner Güte, deinen Spuren folgt Überfluss. In der Steppe prangen die Auen, die Höhen umgürten sich mit Jubel. Die Weiden schmücken sich mit Herden, die Täler hüllen sich in Korn, sie jauchzen und singen.« (Psalm 65, 10–13) Die Dankbarkeit für die Gaben der Erde verbindet sich hier intensiv mit dem Lob ihres Schöpfers, dessen Güte in und durch alle Dinge präsent ist.[24] Dies bedeutet offenkundig nicht, dass der Mensch keinen Beitrag zu ihrer Kultivierung zu leisten hat. Doch die Welt ist für ihn nicht nur Aufgabe, sie ist zuerst und ebenso sehr Gabe. Die hohe Bedeutung der Dankbarkeit im Christentum wird daraus ersichtlich, dass seine zentrale kultische Feier die Eucharistie (griechisch: »Danksagung«) ist. In ihr werden die Gaben der Erde Gott in Dankbarkeit dargebracht, damit sie durch Tod und Auferstehung Jesu Christi gewandelt werden und so die Erde erneuert und die ursprüngliche Gutheit der Schöpfung wiederhergestellt wird (vgl. de Chardin 1990). Nach christlicher Überzeugung ist es den Menschen nicht gegeben, eine vollkommene Welt zu schaffen. Es ist ihm / ihr jedoch aufgetragen, im Hier und Jetzt so zu handeln, dass das menschliche Leben und die Schöpfung erhalten, gefördert und geachtet werden. Ethik und Spiritualität sollen dazu die denkerische und existentielle Grundlagen schaffen.

—Ingeborg G. Gabriel, Universität Wien, Österreich

Literatur

Amery, Carl, *Das Ende der Vorsehung. Die gnadenlosen Folgen des Christentums,* Hamburg, 1972.

Bagir, Zainal Abidin und Martiam, Najiyah, »Islam«, in: *Routledge Handbook of Religion and Ecology,* ed. Willis Jenkins, Mary Evelyn Tucker, John Grim, New York, 2017, S. 79–87.

Baruzzi, Arno, *Einführung in die politische Philosophie der Neuzeit,* Darmstadt, 1993.

Basdekis, Athanasios, *Orthodoxe Kirche und Ökumenische Bewegung. Dokumente – Erklärungen – Berichte,* Frankfurt am Main / Paderborn, 2006.

Basilius von Cäsarea, »Sechste Predigt (Mauriner-Ausgabe Nr. 6): An die Reichen«, in: ders. *Des heiligen Kirchenlehrers Basilius des Großen Bischofs von Cäsarea ausgewählte Schriften,* aus dem Griechischen übersetzt, Band II, München, 1925. Verfügbar unter: http//www.unifr.ch/bkv/kapitel2630–5.htm [3.7.2018].

Beck, Ulrich, *Risikogesellschaft: Auf dem Weg in eine andere Moderne,* 22. Aufl., Frankfurt am Main, 2015.

Broch, Thomas, *Das Problem der Freiheit im Werk von Pierre Teilhard de Chardin,* Mainz, 1977.

Conradie, Ernst M., »An Ecological Critique of Christianity and a Christian Critique of Ecological Destruction«, in: *Routledge Handbook of Religion and Ecology*, ed. Willis Jenkins, Mary Evelyn Tucker, John Grim, New York, 2017, S. 70–78.

ed. Chryssavgis, John, *On Earth as in Heaven. Ecological Vision and Initiatives of Ecumenical Patriarch Bartholomew*, New York, 2012.

de Chardin, Pierre Teilhard, *Der Mensch im Kosmos*, München, 1959 (Original: *Le phénomène humain*, Paris, 1955).

de Chardin, Pierre Teilhard, *Le goût de vivre* (Œuvres, vol. 7), ed. du Seuil, Paris, 1963.

de Chardin, Pierre Teilhard, *Comment je crois* (Œuvres, vol. 10), ed. du Seuil, Paris, 1969.

de Chardin, Pierre Teilhard, »Messe über die Welt«, in: ders., *Das Herz der Materie*, Olten, 1990, S. 120–140.

ed. Demko, Daniela, Elger Bernice S., und Jung Corinna, u. a., *Umweltethik interdisziplinär*, Tübingen, 2016.

Dihle, Albrecht, »Demut«, in: ed. Theodor Klauser, *Reallexikon für Antike und Christentum*, Band III, Stuttgart, 1957, S. 736–778.

Franz von Assisi, *Sämtliche Schriften: Lateinisch / Deutsch*, ed. Dieter Berg, Stuttgart, 2014.

ed. Frevel, Christian, *Biblische Anthropologie: Neue Einsichten aus dem Alten Testament*, Freiburg/Basel/Wien, 2010.

Gabriel, Ingeborg, *Wozu taugt die Tugend? Überlegungen zur Aktualität der Tugendethik*, in: *Die Spannweite des Daseins: Philosophie, Theologie, Psychotherapie und Religionswissenschaft im Gespräch. Festschrift für Augustinus Wucherer-Huldenfeld zum 80. Geburtstag*, ed. Karl Baier, Markus Riedenauer, Wien, 2011, S. 199–219.

ed. Gabriel, Ingeborg und Steinmair-Pösel, Petra, *Gerechtigkeit in einer endlichen Welt. Ökonomie-Wirtschaft-Ethik*, 2. Aufl., Ostfildern, 2014.

Gabriel, Ingeborg, »Ökologie als Frage nach dem neuen Menschen«, in: *Grüne Reformation. Ökologische Theologie*, ed. Michael Biehl, Bernd Kappes, Bärbel Wartenberg-Potter, Hamburg, 2017, S. 83–108.

Habermas, Jürgen, »Israel und Athen. Wem gehört die anamnetische Vernunft?«, in: *Diagnosen zur Zeit*, ed. Johann B. Metz u. a., Düsseldorf, 1994.

Hardmeter Christoph und Ott, Konrad, »Biblische Schöpfungstheologie«, in: *Handbuch Umweltethik*, ed. Ott, Konrad, Dierks, Jan, und Voget-Kleschin, Lieske, Stuttgart, 2016, S. 183–189.

Noah Hariri, Yuval, *Homo Deus: eine Geschichte von Morgen*, München, 2017.

Noah Hariri, Yuval, *21 Lektionen für das 21. Jahrhundert*, München, 2018.

Levinas, Emmanuel, *Humanismus des anderen Menschen*, übersetzt und mit einer Einleitung versehen von Ludwig Wenzler, Hamburg, 1989.

Lintner, Martin, *Der Mensch und das liebe Vieh. Ethische Fragen im Umgang mit Tieren*, Wien, 2017.

ed. Janowski, Bernd und Liess, Kathrin, *Der Mensch im alten Israel: Neue Forschung zur alttestamentlichen Anthropologie*, Freiburg/Basel/Wien, 2009.

Jonas, Hans, *Prinzip Verantwortung*, Frankfurt am Main, 1980 ed. Konradt, Matthias und Schläpfer, Esther, *Anthropologie und Ethik im Frühjudentum und im Neuen Testament: Wechselseitige Wahrnehmungen*, Tübingen, 2014.

ed. Langthaler, Rudolf und Weber, Hubert Philipp, *Evolutionstheorie und Schöpfungsglaube. Neue Perspektiven der Debatte*, Göttingen, 2013.

Leser, Norbert, »Christliche und marxistische Anthropologie«, in: ders., *Jenseits von Marx und Freud*, Wien, 1980.

Moltmann, Jürgen, *Die ökologische Wende in der christlichen Theologie*, in: *Grüne Reformation. Ökologische Theologie*, ed. Biehl, Michael, Kappes, Bernd und Wartenberg-Potter, Bärbel, Hamburg, 2017, S. 27–40.

vgl. Munteanu, Daniel, »Cosmic Liturgy. The Theological Dignity of All Creation as Basis of an Orthodox Ecotheology«, *International Journal of Public Theology* 4 (2010), S. 332–344.

ed. Ott, Konrad, Dierks, Jan und Voget-Kleschin, Lieske, *Handbuch Umweltethik*, Stuttgart, 2016.

Papst Franziskus, Apostolisches Schreiben *Evangelii Gaudium*. Über die Verkündigung des Evangeliums in der Welt von heute, Nr. 51, 2013. Verfügbar unter: http://w2.vatican.va/content/francesco/de/apost_exhortations/documents/papa-francesco_esortazione-ap_20131124_evangelii-gaudium.html [4.1.2019].

Papst Franziskus, Enzyklika *Laudato si'*. Über die Sorge für das gemeinsame Haus, Nr. 114, 2015. Verfügbar unter: http://w2.vatican.va/content/francesco/de/encyclicals/documents/papa-francesco_20150524_enciclica-laudato-si.html [10.12.2018].

Pinker, Steven, *Gewalt: eine neue Geschichte der Menschheit*, 2. Aufl., Frankfurt am Main, 2016.

Pinker, Steven, *Aufklärung jetzt: für Vernunft, Wissenschaft, Humanismus und Fortschritt: eine Verteidigung*, Frankfurt am Main, 2018.

ed. Sarasin, Philipp, *Evolution: ein interdisziplinäres Handbuch*, Stuttgart u.a., 2010.

Schiller, Friedrich, *Gedichte*, 2. Aufl., Berlin, 2013.

Schmidt, Konrad, *Schöpfung im Alten Testament*, in: ed. ders., *Schöpfung*, Tübingen, 2012.

Schor, Juliet, *Wahrer Wohlstand: Mit weniger Arbeit besser leben*, München, 2016.

Schulte, Christoph, *Zimzum. Gott und Weltursprung*, Berlin, 2014.

Silesius, Angelus, *Der cherubinische Wandersmann*, 15. Aufl., Zürich, 2006.

Skidelsky, Robert und Edward, *Wie viel ist genug? Vom Wachstumswahn zu einer Ökonomie des guten Lebens*, München, 2013.

Taylor, Charles, *A Secular Age*, Cambridge, 2007 (deutsch: *Ein säkulares Zeitalter*, Frankfurt am Main, 2009).

Thomas von Aquin, *Summe der Theologie II-I*, Vorwort, hrsg. und übers. v. Joseph Bernhart, Band II, 3. durchgesehene und verbesserte Auflage, Stuttgart, 1985.

Weber, Max, *Gesammelte Aufsätze zur Religionssoziologie*, Band III, Tübingen, 1988.

Zenger, Erich, *Gottes Bogen in den Wolken. Untersuchungen zur Komposition und Theologie der priesterschriftlichen Urgeschichte*, 2. Aufl., Stuttgart, 1987.
Zenger, Erich, u.a., *Einleitung in das Alte Testament*, 6. Aufl., Stuttgart, 2006.

Anmerkungen

[1] So Papst Franziskus in seinem Lehrschreiben von 2015 zur Umweltfrage.
[2] Zur jüdischen Theologie des Zimzum, wonach Gott sich in seiner Freiheit zurücknimmt, um für die Schöpfung Raum zu schaffen, vgl. Schulte (2014).
[3] Der Text der hebräischen Bibel ist (mit kleinen Adaptionen in der Reihenfolge der Bücher und Auslassung der »deuterokanonischen« Bücher: Tobit, Judith, Zusätze LXX bei Ester, 1 und 2 Makkabäer, Weisheit Salomos, Jesus Sirach und Baruch) auch Teil des christlichen Kanons. Dieser besteht somit aus einem alttestamentlichen und einem neutestamentlichen Teil, wobei ersterer aus der Sicht des Lebens, Sterbens und der Auferstehung Jesu Christi neu interpretiert wird. Die Schöpfungsgeschichten sind so Teil der christlichen Bibel. Vgl. Zenger u.a. (2006: 30).
[4] Dies betont Jürgen Moltmann im Ansatz seiner Umweltethik, vgl. Moltmann (2017).
[5] Die Abfolge in der Erschaffung der Geschöpfe steht den Erkenntnissen der Evolutionstheorie des 19. Jhdts. näher als vom Entstehungsdatum her zu erwarten ist. Letztlich orientiert sich die Komposition freilich an theologischen Kriterien, vgl. die noch immer maßgebliche Studie Zengers (1987).
[6] Diese Bestandszusage wird in Genesis 9,1–7 nochmals wiederholt und durch die Zusage erweitert, dass der Mensch nicht von wilden Tieren vernichtet werden wird.
[7] Ersteres zeigt die neuere bibelexegetische Forschung, vgl. Hardmeter und Ott (2016: 183–189); vgl. auch den Überblick bei Ernst M. Conradie (2017) mit weiterer Literatur. Die im deutschen Sprachraum meist zitierte Studie von Carl Amery (1972), führt die Christentumskritik von Lynn White weiter. Der einseitig religionskritische Charakter wird dadurch erhärtet, dass die Wissenschaftsfeindlichkeit des Christentums betont und ihm zum anderen die negativen Folgewirkungen von Wissenschaft und Technik angelastet werden.
[8] Vgl. den Überblick mit weiterer Literatur von Zainal Abidin Bagir (2017: 81 f.).
[9] Vgl. den umfassenden Überblick über die Entwicklung der westlichen Kultur in Bezug auf ihr Naturverhältnis bei Taylor (2007). Erst über den Deismus, der die Schöpfung durchaus wertschätzend nach Art einer Maschine denkt, die des Eingreifens Gottes nicht mehr bedarf, wird die Subjekt-Objekt-Dichotomie der Moderne möglich.
[10] Die wichtigsten Schöpfungspsalmen sind Psalm 8, 19 und 104. Wie Schmidt hervorhebt, »führt aber auch der Gesamtduktus des Psalters im Bereich seines ›fünften Buchs‹ (Ps 107–150, […]) auf eine profilierte Schöpfungstheologie hin«, vgl. Schmidt (2012: 101).
[11] Angelus Silesius (2006); Franz von Assisi (2014).
[12] Habermas knüpft damit an die von Max Horkheimer und Theodor Adorno bereits 1944 formulierte These von der Dialektik der Aufklärung an.
[13] Als Beispiel sei hier die Monotheismuskritik im Gedicht Friedrich Schillers, *Die Götter Griechenlands* genannt. Hier heißt es in einer der Verszeilen: »Einen zu be-

reichern unter allen, mußte diese Götterwelt vergehn« (Schiller 2013: 127). Sie stellt somit einen Vorläufer jener Kritik dar, die – siehe oben – seit den 1960er Jahren geäußert wird, vgl. Fn 8.

[14] »Le *progrès – être plus*, ou bien il ne signifie rien«. Pierre Teilhard de Chardin, *Sur le Progrès* (unveröffentlicht), zit. in Broch (1977).

[15] Vgl. dazu Gabriel (2017), sowie die Beiträge in Gabriel und Steinmair-Pösel (2014).

[16] Vgl. dazu mehrere Beiträge in Ott, Dierks und Voget-Kleschin (2016), besonders 322–329, 330–334, 335–342, 343–347, 348–352, 353–360; sowie Demko, Elger, Jung u.a. (2016).

[17] Nicht eingegangen werden kann hier auf den praktischen Beitrag der Kirchen zur Umweltethik und -spiritualität, vgl. im Blick auf die Orthodoxie Basdekis (2006); weiters Chryssavgis (2012).

[18] *Humilitas* eignet sich als Begriff offenkundig besser als das vielfachen Missverständlichkeiten ausgesetzte deutsche Äquivalent der Demut.

[19] Dies wäre u.A. für eine Tierethik von beachtlicher Bedeutung, vgl. Lintner (2017).

[20] So das Dokument der Bischofssynode der Katholischen Kirche *De iustitia in mundo* von 1971 (!): »Die Nachfrage der wohlhabenderen Länder nach Rohstoffen und Energie (wie auch die schädliche Wirkung ihrer Abfälle auf Atmosphäre und Ozeane) hat ein solches Ausmaß erreicht, dass die wesentlichen Voraussetzungen des Lebens auf dieser Erde wie Luft und Wasser unwiederbringlich geschädigt würden, wenn die Höhe des Verbrauches, der Grad der Verschmutzung und die Schnelligkeit des Wachstums bei der gesamten Menschheit Platz greifen würden«. Nr. 11. Verfügbar unter: http://www.iupax.at/fileadmin/documents/pdf_soziallehre/1971-weltbischofssynode-de-iustitia-in-mundo.pdf [25.12.2018].

[21] Auch die Konsumforschung sieht in einer Verringerung des Konsums den Grund für mehr Freiheit und Zufriedenheit, vgl. u.A. Schor (2016).

[22] Vgl. aus der Fülle der Literatur Robert und Edward Skidelsky (2013).

[23] Dieser ist gegenwärtig für etwa 1/3 aller CO_2 Emissionen weltweit verantwortlich.

[24] Dies war eine Grundeinsicht der frühen Theologie. Vor allem Maximus Confessor (580–662) vertrat eine derartige kosmische Theologie, vgl. Munteanu (2010).

Rituals as Environmental Skills: Inhabiting Place, Fabricating Meaning, Enhancing Morality

Do rituals have a place in our perceiving, thinking and acting within nature? Do they have in general a place in modern life, and, in particular, a place and function in environmentalism and environmental ethic? If so, might rituals offer unique social skills to »make-oneself-at-home« and to assist the just and sustainable flourishing of »Earth, our home« (www.earthcharter.org, 18 September 2006) in a world of constant homelessness and increasing anthropogenic dangerous environmental-and-climatic devastation? Is there any significance in rituals and processes of ritualisation for our perception of, thinking about and acting in, for and within the environment?

This chapter responds to these questions positively. Rituals are by no means regarded as ultimately pointless to be fenced in the ethnographical museum, but rather treated as essential for social life in general. In such a lens, rituals appear as specific skills and places where both meaning and belonging can emerge. As such it might be worth to mine them deeper with regard to their potential power for and within environmental ethics in an inter- and transcultural key.

In the following I will depart from the common understanding of rituals in the transdisciplinary field of ritual studies, and hereby especially emphasise the discussion of rituals in the context of environmentalism. The discussion though seems still limited in spite of its unexploited potentials, and I can therefore in the frame of this text only offer preliminary reflections on some selected subthemes: ritualised fabrication of meaning, ultimate sacred postulates, inhabiting a place, spiritual and environmental activism among the Mi'kmaq, machine fetishism, and rituals in urban space.

1 Rite, Ritual and Ritualisation—Meaning, Power, Wisdom and the Sacred

Resonating with the well-established discussions in ritual studies and ritual theory, rituals are regarded as sociocultural mediums that invoke the ordered relationships between human beings and non-immediate sources of power, authority and value. Rituals represent sociocultural mediums constructed of »tradition, exigency, and self-expression« (Bell 1997: xi). Following Catherine Bell, they are enabling people to embody assumptions about their place in a larger order of things. »Ritualisations« should not be taken for rituals but are those actions that transform a practice into a ritual. »Rites«, in contrast, are ceremonial acts that take place mostly in explicitly religious contexts. They are further grouped in different categories of practices such as rites of passage, communal or individual rites, or also pilgrimages. One should also keep in mind that ritual scholars do not follow a clear consensus about definitions and even avoid pragmatically all too strict limitations of their central terms (cf. Grimes 2014: 189–90).

Social processes can thus be studied in the new light of the young field of ritual studies, mainly developed in the frame of cultural and religious studies. The lens of looking at social life as rites, rituals and ritualisations offers exciting possibilities to combine the analysis of actions and ideas. Looking at social life in the ritual lens also allows one to become more aware about the interconnection of perception, thinking and acting. Following Tim Ingold, thinking and acting are intimately interwoven but they are both rooted continuously in our perceptions of the environment (Ingold 2000: 166 ff.). Consequently, the impacts of rituals on thinking and acting should therefore be emphasised through their capacity to interact with our perception. Performing a ritual affects the understanding of oneself within a community and surrounding; it permeates the perception of the self and its embeddedness and opens new horizons that can persist also after its performance.

Such a multifaceted view can also enrich environmental ethics in a fruitful way. Sights, values, worldviews, moralities and practices appear in this perspective as closely interwoven, and they are not in the same way exposed to philosophical systems' well-known risk of fragmenting and separating practice from value, seeing from acting, and subject from object.

Rituals are not necessarily embedded in religion, even if religions substantially draw on ritual practices and rites in their unfolding. Rituals can fabricate meaning, but they can also appear meaningless. Not dependent on the ambivalence of ordinary life, rituals can—as basic social acts—offer a kind of safe language for performative effects. The marriage ritual's changing vows gets you for example married even if you are not in love at all with the other. Once a ritual has been performed under the right conditions it develops its own law and power.

Rituals are able to create and carry a community, and they produce what one might call specific ritual places that are able also to impact on a broader surrounding. Rituals are able to transform places into sacred places either for a short or a long time. My introductory phrase above about rituals as *skills and places* was formulated to express this place-shaping and place-transforming capacity.

The older assumption that religion—and rituals with them—would decline in the process of modernisation has, as we know, not come true. Not every ritual represents necessarily a confessional practice of faith, but religious practices can scarcely develop without ritual performances. In fact, religions can be analysed as ritualisations that depend on and reshape space. Whatever concept of religion one prefers, the foundational code of religious processes approaches reality as a given spatiality in connection with the acting spirits, gods, or God, and rituals are locating faith at place. The interaction of religion and ritual and its scholarly discussion though is all too complex for to be mapped in detail here (cf. Insoll 2011).

Today processes of so-called secularisation and re-sacralisation are flourishing at the same time, more or less interrelated, and they increase in the same common modern global space, though in different cultural forms, within the one common world system. Nevertheless, the overarching theories about secularisation and sacralisation seem to have lost their capacity as grand narratives to explain the overwhelming diversity and dynamic characterising »The Changing World Religion Map« (Brunn 2015), where one for example in the field of nature, ethics and the environment can clearly observe new modes of searching for the sacred (Buttimer 2015).

Processes of ritualisations have furthermore been identified and analysed in growing different social spheres. With regard to rituals, which earlier were regarded only as instruments within a cultural system, one can surprisingly see a development where ritual practices themselves are regarded as objects for belief, rather than ideologies or images. This means that partaking in a ritual performance is believed to be more important than to rationally accept a set of doctrines about God, world and man/woman or to adore a specific object of divine representation. Rites, rituals and processes of ritualization are, as Roy R. Rappaport, Ronald Grimes and Catherine Bell have shown, at the daily heart of fabrications of meaning in social action.

Another interesting capacity of ritual lies in its normative power. Rappaport has in his famous ecology of religion shown how rituals work as »homoeostatic«, that is as aesthetically normative regulations of social interactions with nature (Rappaport 1968 and Rappaport 1999: 411). Due to ritual practices, humans preserve images and norms about how to interact with hunting, farming and modes of survival. Rituals can work as normative practices which regulate human ecology, in traditional communities as well as in modern societies.

In his later work, Rappaport regards ritualisations (the expression of belief and meaning in ritual practices) as the essential way of fabricating meaning. He analyses modern society by distinguishing between scientific and religious interpretations of reality. The ritual fabrication takes hereby place in contradiction with the scientific rational production of knowledge, or what he calls the »epistemologies of discovery.« In contrast to the boisterous rationalists, who call for a sharp separation of knowledge and belief, Rappaport pleads for a necessary synthesis of both (Rappaport 1999: 451 ff.). For him, the basic tension of human life lies between constructing meaning for our environment and the epistemologies of discovery that aim to explain the laws of the same environment.

Even if Rappaport and others in their ritual studies have shown how environmental regulations take place as efficient practices in the ritualization of a culture's agency with natural resources, it is controversial in scholars' internal discussion how this works and whether rituals *should* function as environmental regulators or in fact are *doing* it. If ritualization is regarded as a basic social act, it represents a »cosmo-topic« practice where the human cosmos is established performatively as a »cosmo-sized place«, a universe at a local place (Grimes 2003: 44).

Also, in climate change this tension between the rational observation of global and local environmental change and the fabrication of its meaning is detectable. But scientific explanation should in my view not override the production of local knowledge and it should further include also ritually fabricated meaning, religious or otherwise spiritual meaning. Following Rappaport, both need to progress and interact. Such a synthesis can turn into the well-needed wisdom that, according to Nicholas Maxwell, is necessary to decide how to apply the products of rational knowledge. In his criticisms of the philosophy of knowledge, which does not take into account the living beings' self-interests and interplay and which only seeks simple rationalisations of selected phenomena, Maxwell argues that we should replace knowledge with wisdom inquiry (Maxwell 1984). A philosophy of wisdom is for him a truth-seeking discourse that strives to negotiate the criteria of why we should act or abstain. With Maxwell's help, where rationality and wisdom are not equal but placed over and under each other, we can thus sharpen Rappaport's plea for a synthesis and regard rationality as a presupposition for wisdom inquiry and ritual as an experimental method to explore its practice.

Applying Grimes' insight one can further trace a close entanglement of narratives about the change of climate and our common future and the ritualization of a climatically cosmo-sized place that is depicted in new colours, and ask if how such cosmo-sizing ritualization might contribute to the production of new environmental wisdom. How can one express the presence of the whole universe at one particular place in rituals?

2 Ultimate Sacred Postulates, Ultimate Concerns and Empowering Gods

Rappaport furthermore emphasises in his later work the significance of »ultimate sacred postulates« (Rappaport 1999: chapter 9), that is, the central, normative presumptions that determine the value of actions in light of what we can describe as ultimate concerns. No matter how much these are constructed, humans experience and regard them as real and sacrosanct. One can disregard them but not falsify them, as they exist beyond empirical verification. Also, technical and rational ideologies are in my view constantly hiding such ultimate sacred postulates. For example, technology, as well as social engineering, is fueled by the *belief* in total controllability and achievability, whereby one can control humans and natural processes through artefacts and machines. Inspired by Rappaport, one can here on the one hand explore the sacred postulates and hidden driving forces within machine fetishism but at the same time also empirically investigate the social construction and significance of machine fetishism.

Another example can be found in the ultimate sacred postulate about money, and the belief in the final value that is placed on monetary accumulation and leads to the worldview and practice of applied mammonism. In our discourse on environmental ethics the awareness about such implicit ultimate sacred postulates in presumed rational arguments can help to discurtain problematic claims of validity.

How can one, for example, believe in the constructive power of geoengineering in general, which attempts to solve problems that have been produced by other forms of engineering, particularly fossil emission-based technology? Why should faith in machines save us from problems caused by machines? And how can processes of ritualization catalyse or resist such fabrication of meaning? What kind of a cultural system and process of ritualization do we need for example in the sphere of engineering if we want to enhance life at depth rather than to govern and rule it for particular interests of some of us? Can also scientists (and engineers) transform into more than simply »a family of innovative dwarfs that can be hired for everything« (Brecht 1939–55: Bild 14)? How can ritualisations assist such a transformation (for example in higher education institutions)?

Surprisingly enough, the insight from ritual studies about the normative power of ultimate sacred postulates converges with a central insight that has for a long time also been at the core for theologians in the Protestant tradition. In analogy to Rappaport's thinking, the influential philosopher of religion and theologian Paul Tillich coined his technical term of »ultimate concern« in the 1950s. On the one side, this concern characterised for him the essence of religious attitudes and demands that all other concerns are sacrificed. On the other hand, Tillich circumscribes Christian faith as ultimate concern, a belief that transcends rational and other expressions in an ecstatic passion for the ultimate (Tillich 1957: 8–9).

Ultimate concerns are not only at work among believers in explicitly religious confessions but they characterise a deeper existential skill to form some kind of a hierarchy of normative ideas in worldviews too. Tillich continues hereby to walk on a path that earlier had been ploughed by Martin Luther, who in his »Large Catechism« has circumscribed »god« as that »what you set your heart on and repose your trust in« (Luther: First part, First command). In Tillich's concept god is »the name for that which concerns man ultimately« (Tillich 1951: Volume I, 211). Turning back to our discussions about rituals we can conclude that their normative power anchors in ultimate concerns, expressed in ultimate sacred postulates, as well as in the imagination of gods to whom we set our heart, trust and rely upon. In the context or discussing environmentally relevant rituals one can learn from all these three, Rappaport, Tillich and Luther, to become aware about the normative power of overarching, and often hierarchically structured, ideas and values that can unfold an immanent autonomous power to govern social and ritual life within nature.

From theologian John B. Cobb Jr. and environmental economist Herman Daly's classical work on the »Common Good« one can in the same plough of thinking learn to regard also theocentrism (not simply as a source of all premodern monistic top-down governing evil) but as a significant »check against idolatry« (Daly/Cobb 1994: 401). Efficiently belief in the power of one God makes it possible and even necessary to question and to relativise all demands for power from other gods and to refuse these if they aim at conquering the top of ultimate concerns and sacred postulates. The idea of the one God's universal power works in this sense as an idea of supernatural and superhuman power in order to limit and encapsulate human power claims and to keep these open and negotiable, as Cobb has convincingly argued (Daly/Cobb 1994: 401–404).

To whom do we set our trust? And in what kinds of rituals do we express our faith? Which gods are allowed to empower us and to whom/what are we setting our heart? How do we lend power to rituals so that these can unfold their own authority and power? What kind of beliefs and connected ritualisations are supporting or countervaluing existing power constellations? Obviously both a spiritual and ritual perspective on questions like these can throw almost some light for unmasking hidden and implicit, life-enhancing as well as life-devastating, sociocultural driving forces.

3 Ritually Re-earthing and Inhabiting a Place

One of the few well thought-through texts on the theme of this chapter has been written by leading ritual scholar Ronald Grimes. He departs from the observation that few people on the one side would consider rites as effective means to save the environmentally threatened planet but that one can follow on the other side an emergence of groups and individuals for whom it is obvious that rituals are offering one of several significant responses to the environmental conundrum (Grimes 2003: 31).

Grimes refers to different examples such as the emergence of the »Council of All Beings«—a communal ritual, founded in Australia in 1985, in which participants step aside from their human identity and speak on behalf of another life-form—with a series of »re-earthing« rituals and workshops where humans are trained in carrying animal masks and where participants' deepened connection with Earth and its creatures shall lead to a better morality and political commitment for the sake of the planet. Another example is found among Buddhist monks in Thailand who have created environmentally oriented rites and applied these in their struggle against clear-cutting land bulldozers driven by corporations' profit interests. Also, composers and artists from the 60s have developed ritualised musical performance rites onwards.

While environmentalists are, as we have learned recently in the sociology of religion and environmentalism, approaching nature in a neo-animist or pan-spiritist way, where living creatures and systems are regarded as spiritually animated (Harvey 2006: 212), Grimes states aptly that ritual responding to environmental problems is »not typical of the religious mainstream of the Euroamerican West« (Grimes 2003: 33). Looking at the more recent developments in the Christian ecumenical movement as well as in local churches, it is obvious that not only rational (eco-)theological and ethical argumentation is at the core but that also other modes of mobilising believers inside and contribution to public debates outside are intensified. Liturgical renewal and aesthetical expressions such as music, drama, visual arts, architecture and design play herein central roles. Rational and ethical argumentation takes place side by side with ritual and aesthetical creativity, with the same intensity that one also can observe in constantly growing social environmental-and-justice-related movements, for example in the annual »World Social Forum.« Exciting countervailing power is produced in approaches such as the »Dark Mountain Project« where stories that prevent us from seeing ecological, social and cultural unravelling are contrasted to other stories of help and hope (Dark Mountain).

To the question if rituals might be good for the environment Grimes responds positively:

> For attitudes to become definitive they must be cultivated by practice. And the name for sustained, value-laden attitude practice is ritual. In ritualizing, human beings discover, then embody and cultivate their worldviews, attitudes, and ethics. Rites are not only about confirming views that people already hold but also about divining ways to behave (Grimes 2003: 33 f.).

Grimes regards rituals as both cultural and natural. According to him all social behaviour is »not only ritualized but necessarily ritualized«. Ritualization rather represents an essential function of our brain and nervous system, and therefore both a function of our biological and our cultural constitution (Grimes 2003: 36). Convincingly Grimes both acknowledges and criticises biogenetic structuralists who focus mainly on the biological anchorage of rituals and he defends also the need for an ethical critique of ritualistic means. What I herein deeply appreciate is that one in such a perspective can interconnect ritual and ethics in a differentiated reciprocally critical way, which in my view seems inevitable for not to fall into simplistic gaps.

In a detailed discussion with Rappaport's earlier dominant and influential approach Grimes acknowledges the ecological anthropologist's view that ritual is the basic social act but criticises fittingly his inability to do justice to rituals' creativity by focussing all too much on their function to preserve the existing order (2003: 40). Ritual for Rappaport, in Grimes' eyes, mainly creates social and ecologic order and its most essential function is to preserve this order for the sake of the planet. Provokingly Grimes sums it up:

> For Rappaport, ritual is essential to planetary survival not because God created it but because it created God. (Grimes 2003: 41).

Rappaport takes his approach even further, into some kind of a mystical transfiguration (my term) by regarding humanity itself as that part of the world that can think about itself, so that ritual performance becomes not merely a means for humans to express their relation to the environment but ritualising becomes »the way the world itself tries to ensure its own persistence« (Grimes 2003: 43). While rituals for Rappaport are mainly conservative and preserving the world's persistence, Grimes is able to perceive and analyse rituals in a more differentiated way. Rituals can in that sense offer creative social skills for to »enhance adaptability and thus the longevity of the human species on the planet earth«, and they can also in deep ritualised life render »participants one-dimensional, stereotyped, and inflexible—in short, maladaptive.« (Grimes 2003: 43)

Following Grimes in his conclusion, this ambiguity should in my view rather be regarded as a source of creativity for socio-environmental transformation than a hindrance. Rituals allow, as Grimes so aptly formulates, »to discover ways of inhabiting a place« (Grimes 2003: 44). Rituals help people to figure out, divine and even construct a cosmos. In order to make ourselves at home on Earth, we do need to interrelate the local and the cosmic, at every place anew, and rituals are nurturing this process. A cosmos is not merely an empty everywhere, as Grimes makes us aware about, but it has to be loaded and filled with meaning. Rituals contribute in such a lens to establish cosmosized places and they allow to both act for the preservation, protection and transformation of environmentally threatened life.

Till now, I am talking mostly in general about what rituals might and maybe also ought to do, but I almost hope to have awakened the reader's curiosity about the potential of rituals in their own means for the best of environmental commitment before we shortly will take step closer to their practice. Rituals can make us dance, sing, paint, taste, play-all activities to explore and discover new and old ways of inhabiting a place. Especially environmentally threatened places and spaces can—by the means of ritual performance—appear in new colours, sounds and shapes; rituals do not solve problems but they allow to perceive them deeper and to respond differentiated, no more or less. Ethics therewith becomes able to turn into what I have called »aesth/ethics« (Bergmann 2006, 2010, 2014), that is the embodied perception about what is morally demanding at depth. The slash was once inspired by Wolfgang Welsch's invented term »Ästhe/tik« (Welsch 1996: 108) that however depicts only those parts of aesthetics which of one own's accord include ethical moments, while »aesth/ethics« in my sense makes us aware about the aesthetical foundation of ethics in general. Ecorituals are in such a perspective not only driven by or applying values, moralities and ideologies, but they are also able to deepen and widen perceptions, to provoke, and to critically question and transform common views and methods. How such a reciprocally critical interconnection between ethic and ritual can take place in practice and what it can teach us, seems still to be a wide-open and all too less investigated field of research. Nevertheless, I would like to offer some selected insights into the field of such exciting interaction.

4 Ritual Power in Spiritual and Environmental Activism

A move to the Americas and the land of the Mi'kmaq Indians, living at Cape Breton Island, Nova Scotia in the North East of Canada, takes us closer to the field of ritual power in the context of environmentalism.

Anne-Christine Hornborg, a scholar of religion, has in her ritual studies (Hornborg 2001) carefully investigated the field of environmental protests among the Mi'kmaqs, where rituals are used to prevent exploitation of places and natural resources. Knowledge from science, that is mainly environmental science about the impacts of mining, has in this case been combined with expressions of religion and spirituality. In a synthesis of what Hornborg depicts as »integrative science« rational knowledge, religion and environmentalism have been compressed in meaning conveying practices and rituals. What the Mi'kmaqs regard and respect as their sacred mountains was violated brutally by mining companies' and national bodies' profit-driven and energy hungry interests. Hornborg's extensive studies explore excitingly the power of ritual performance in this context. Traditional rites as well as creatively invented new rituals offered an excitingly creative social process, where also modes of thinking in established environmental ethics were combined with findings in environmental science. While Grimes' and Rappaport's discussions still approach ecorituals as a marginal and potential force, Hornborg's study offers evidence for the emergence of a new kind of what scholars depict as »spiritual activism.« It furthermore clearly shows the difference and collision between bureaucratic language and ritual performance in the struggle, where a ceremony successfully could resist verbal argumentation. Spiritual activism can specifically also turn into spiritual eco-activism, in the land of the Mi'kmaq as well as in other spheres, which we recently discussed in a specific international conference on »Faith Communities and Environmental Activism« in Edinburgh in May 2017 (Faith activism conference).

Also, in other contexts, and even in not explicitly religious contexts, one can observe the growth of such activism, for example in how a healthy life style increases its value and leads to an increasing demand for ecologic food. Food practices are not only longer just ordinary but driven by such activist values where many avoid animal food because of ideological and ecological reasons, where animal care is debated and where animals in general change their significance for humans. Eating has in such a sense become rather an existential issue and for some also a somehow magic performance. What we do allow to go into our body also impacts our identity in an animated surrounding and universe. An increasing ritualization of food producing and sharing practices can consequently be observed. After her Mi'kmaq studies Hornborg (2005, 2009) continued to critically analyse the process of ritualization and explored the health industry and individuals with regard to »health rituals« (Hornborg 2009). Such an approach establishes undoubtedly an ambitious new field of relevance also for environmental ethics, a field where the marketing and selling of nature and health take place in parallel with each other.

The meaning of place, that Grimes had emphasised so aptly, was at focus also in the Mi'kmaq's resistance against the companies' and national economic interests, in this case around and at the Kluskap's Mountain on Cape Breton, and its cave. Already the Canadian name of the mountain expressed in the indigenous view an oppression, as it was the non-native smuggler and moonshiner Keller, who lived on this mountain, who also gave the official name, Kelly's mountain, to the place. As a part of their struggle, embedded in a long history of centuries of colonisation and oppression, the Native Americans renamed the place as Kluskap Mountain, asserted this as the home of their culture hero Kluskap and the place where he was expected to return to his people.

During the 1990s, the Mi'kmaq fought a successful battle to prevent the establishment of a super-quarry on this mountain, and employed constantly rituals in their protests, a process that appeared to be a successful way of resistance, within their own groups as well as within the public discourse that accompanied the events. Referring to inviolable values like tradition and sacred places, practiced in rituals in combination with ethical and scientific perspectives, and also in conscious and careful integration with Catholic celebrations, feasts, saints and liturgies, where one could gather native and non-native believers, produced effective results. Also, Catholicism turned in this context into a common ritual practice of resistance that Hornborg analyses in the tension between »power play« and »reconciliary hegemony« (Hornborg 2005: 103fff.). »Today it does not seem that the plans for a quarry will be realized«, Hornborg reports (Hornborg 2008: 8), even if many will play down the significance of the spiritual activism for the sake of profitability and stronger technological demands from the state. The mountain thereby remains to be a sacred place protected by, among other, also rituals.

While Rappaport, in Hornberg's view, departed from the liturgical rite as the raw model of rituals in general, the new spiritual activism operates with a different understanding and relation to place. While older rites and also for example the Mi'kmaqs in their struggle could apply rituals for to protect, to defend and to preserve the traditional power of place, new rituals »must attract a more or less placeless, detraditionalised modern human, and this makes the prerequisite for the ritual performance different than for the traditional liturgical rite« (Hornborg 2009: 111). Health, for example, can in such new activism turn into a commodity to be traded and economised.

Another difference lies in the distinction between community and individual. While older traditional, also environmentally relevant, rites were mainly community-based, new rituals are individually cantered; they are further prepared for constant change according to the needs of the participating persons. New ritualised practices are according to Hornborg also developing new relations to specific places which they discover and even produce, such as health centres, fitness centres, and alleged green nature zones (for bodily and spiritual well-being and comfort). Many of these pretended new rituals have their roots in Eastern Asian traditions and practices which they transfer, fragmentise and transcontextualise to the west inconsiderate of their original context and function. The individual body is hereby ritualised and treated as a central object. Critically, Hornborg discusses how rites can turn into commodities that serve exclusive financial interests. Ritualisations can insofar not only serve as contributors to save our common »Earth as our home« but also serve as money-makers for some of us. Cultural traditions are exploited in that way and the old colonial cycle can continue in new forms. Ritualisations can in this sense function as instruments of neocolonialization where ritual raw material from one original context is commoditised, transported and sold back to new actors for a cheap price, or better for nothing that can benefit the cultural practice's original inventors and guardians of rituals that once might have served their practitioners and others' life.

Learning from Hornborg's studies in the Mi'kmaq land, the health industry and the ecofood sphere, one cannot uncritically set one's trust only to the power of rituals themselves for to save the planet but need to be cautiously and critically aware about the rituals' cultural context, driving interests and power constellations. Critical ethical questioning is therefore more than needed also in the context of ritual performance and its background warrants.

Furthermore, one can learn that obviously a well-working deep synthesis of rational, ethical, aesthetical and ritual dimensions seems to be stronger than only rational or ethical argumentation if one wants to catalyse environmentally life-enhancing social processes. Also, rituals are, as we have seen, deeply in the need of self-critical awareness. Rituals can without doubt serve as new cultural resources but they are by no means ethically non-ambivalent. Discovering new modes of inhabiting places and developing forms of co-existence in environmental spaces demand ritual, conscious, spiritual and aesth/ethical awareness in a well- functioning synthesis.

Rituals can, as we see, serve as significant and unique cultural skills to respond to crises. However, they cannot be taken as the only skills but seem to function best in cooperation with other skills such as rationality, ethical consciousness, bodily awareness and aesthetic creativity with regard to the suffering of one's neighbours both in human and in non-human spheres of life, both nearby and far away. Without doubt, rituals can hereby contribute with unique skills to respond to an increasing amount of suffering in climate-related processes of change, and to the social and cultural re-ordering of sustainable and just human life on Earth. I have difficulties to imagine how the urgently needed mobilisations for the »Great Transformation« ahead of us (Schneidewind 2018) can take place without the power of traditional and creative new ritual performance, but I have also my doubts, with Grimes and Hornborg, about relying solely on the power of rituals, as these seem to stay, in analogy to many other human skills, in a constant danger of misuse and commodification.

5 Machine Fetishism and Rituals of *Verblendung*

Another way of exploring our theme about the interconnection of ritual and environmental ethics takes us to the discourse about technology and what engineering and machines, and the almost religious adoration of technical systems, do to us, to our co-creatures and to the environment. Following my earlier work on machine fetishism, environmental animism and the potential of a reconstruction of beliefs in the Spirit, including Christian faith in the Holy Spirit as a Giver of (sustainable) Life (Bergmann 2012, 2015), I would here shortly like to emphasise anthropologist and human ecologist Alf Hornborg's critique of our contemporary technocratic system in the lens of ritual.

According to Hornborg, modern technology can undoubtedly, together with global turbocapitalism in unequal and unjust social systems, be held responsible for most of the accelerating dangerous climatic-and-environmental devastation. It is supposed to work as some kind of magic. What is the difference between the efficacy of magic and the efficacy of the machine? Hornborg explores this analogy thought provokingly (Hornborg 2014) and relates it to the question of how pre-modern sacred and ritual power could be transformed into modern economic and technological power.

Industrial technology as well as theocratic rituals, for example in ancient Peru, follow according to the anthropologist the same cultural dynamic. While symbolic objects such as spiritually powerful shells in Peru could offer evidence for the successful function of ritual communication with the empire's divine father, the Sun, and thereby guarantee agricultural fertility, also modern society is dependent on such similar »pivotal exchange rates« as for example oil prices. The transformation of imports into work has in modernity been locally objectified into technology, and machines are in themselves manifestations of global exchange rates.

> No less than ritual, machines mystify us by pretending to be productive independently of exchange rates. In modern capitalism, however, the mystified exchanges have become even more opaque, and the magic agency of fetishized objects has become compelling in completely new ways (Hornborg 2013: 256).

Modern power relations are developed in analogy to premodern ritually negotiated power, where social elites are able to extract symbolically controlled obedience and labour energy from all the many human beings who provide them with the means of asserting these demands.

Rituals have, as we have learnt aptly from Rappaport, even a deep conservative and power preserving potential function. In our own times also the promises of continued economic and technological growth, and even of global sustainable development, are dependent on such ritualised and spiritually fabricated meaning. With regard to the propagandistic myth of eternal growth, at present fruitfully resisted by the post-growth movement, they are underpinned by more or less reliable quasi-scientific explanations such as neo-liberal reductionist economy models and other de-contextualised anthropocentric science such as the theory of ecosystem services (as if all nature only would exist for the sake of serving man). In my view such an unholy alliance of ritual, quasi-scientific and culturally normative symbolic values, offers a gigantic process of ›blending-off‹ (German *Verblendung*) that environmental ethics, gladly in cooperation with ritual studies, needs to unmask and practically deconstruct.

In the same way as the Mi'kmaq mentioned above, other environmentally committed groups are searching new modes of inhabiting and transforming occupied places with the help of rituals, also the processes of common *Verblendung* and technocratic seduction demand their own rituals such as selling one's own ritualised healthy body, identifying oneself rhetorically and practically as a confessing motorised »car believer« or handing over one's autonomy to the contingent games of fortune play, gambling or dreaming of a profit on the stock market. A manifold of contemporary social processes can thereby be analysed in the lens of ritualization. Rituals serve different interests, and they need to be constantly critically investigated with regard to their normative implications for others.

6 Rituals in Urban Space, Spatial Justice, Movement, and Aesth/ethics

Finally, I would like to turn to another highly exciting field: how rituals work in urban spaces. Environmental humanities and sciences have increasingly emphasised the ambivalent significance of a rapidly growing settlement of human beings in urban spaces, and the question of how spatial and environmental justice can be nurtured in the global »postmetropolis« (Soja 2000) has consequently moved to the top of the environmental agenda. For environmental ethics, the turn to the spatial is crucial. As a majority of the planet's population lives in an urban environment, we are facing a significant need to reflect deeper about the spatiality of environmental ethics. Issues of energy use, housing, water ethos, urban planning, mobility, life style, and consumerism need to be thought through anew by also regarding the impact of built environments. My short section here can only offer glimpses but hopefully encourage further research in the field of environmentally based third space studies, that is to develop studies of physical *and* symbolic space as »lived space« (cf. Soja 1996 and Bergmann 2014, 49–70).

Also, in urban space one can observe how rituals contribute to the fabrication of meaning alongside the rational and scientific approaching of reality. Rituals regulate spatial behaviour and imagination in a normative way. How is this true in late modern urban space today?

Even if our understanding of a democratic society persuades us that all places are open for everyone, reality does not offer evidence. In a similar way, as temples in premodern times have had a hierarchy of spheres which were more or less open to all or strictly limited to religious elites, also the late modern city offers what I would call *maps of avoidance*. Some areas will be avoided or only accessed with very specific purposes, such as banks, embassies, or business buildings. Others are only approached with purposes such as schools, hospitals, shops, and theatres. Only few buildings are truly open public places such as squares, botanic gardens, museums and railway stations. An exciting example of a strongly ritualised space is the modern airport, where rites of passages take place several times and where a strict regime of surveillance and control is surveying all movements in the sharply drawn map's territory. With regard to environmental ethics, usually green spaces in cities, such as parks, cemeteries, alleys, squares, solitary trees, and playgrounds, offer a rich panorama of environmental conflicts. But also, styles of architecture, maps of mobility, and the planning of social housing offer a dynamic field for collisions between different understandings of dwelling and what characterises a city worth to live.

An exciting example for how an artistic process can break out of the segregated urban map is found in the Swedish city of Malmö a dark autumn week. The short-term construction of a place for open ritualization can exemplify what in recent years has been tried out also in many other cities. One week after Halloween the cultural association *Rárika* arranged in 2006 a successful *Festival of the Dead* in order to counteract the asymmetry of life and death in modern culture. A couple of events in the darkening Nordic autumn took place and encouraged inhabitants to share their practical expressions of the sacred, or better: of what one regarded as sacred. Among other places especially *The altar of the dead* touched me, with many others, deeply. Some kind of an open cave was designed in the city's central park, and it was installed without any signs referring to a specific religious confession. Frequently many inhabitants used it for to light candles, to contemplate, and to experiment with different forms of ritualization. Also, artefacts in this context attained a new distinction as »things as links between person and place« (Peterson 2009). The event in Malmö offered in this way a vibrant contribution to rethink what »ars moriendi« (the art of dying) means within environmental ethics in a technocratic society where death is allegedly eliminated and the dying are pushed to the margins (cf. Altner 1981).

Another interesting observation is about movement. Bodily movement as well as emotional motion seems to be crucial in every ritual practice. Rituals of birth and death accompany the new-born and the dead neighbour. While the first moves into existence, the other moves out this world into another one, and rituals are mirroring both these movements. Movement in itself is central in for example rites of passage or pilgrimages, while other rituals connect to the rhythm and circle of natural seasons. A ritual turns space into symbolic space and time into symbolic time. Movements according to specific spiritual references sacralise space and time. While some initiated are allowed to move into specific spatial spheres, others have to stay outside or in limited spheres. While some places are open for all, others have to be avoided in order to keep the order alive. The ritual that connects the community to the invisible powers achieves spatial order. Such a relation is central for social harmony, spatial order and the meaningfulness of a community. Also, in modern and late modern times ritual practice as movement works in this way. How is ritual linked to the mobility of urban space? Could we regard »the ethics of mobilities« (Bergmann and Sager 2008) as some kind of a ritualising practice, where ritualised movement regulates urban mobility? How are modes of moving ritualised in walking, biking, sailing or in motorised high-emission transport?

Reflecting over rituals of the sacred in the modern city one would need to follow a double direction of impacts. How does urban space affect ritualization, and what do rituals mean for lived space? It seems obvious that the choice of place is essential for the development of a ritual. Actions create some kind of what I have called a »ritual place« (Bergmann 2012a/2014) and decisions about the choice of place are part of a broader negotiation. A procession, for example, moves from one (sacred) place to the other, or it carries the charisma of the sacred from the inner to the outer. A procession also changes the surroundings where it moves along. As long as the ritual endures, a place can no longer be used for transport or trading; also, the remembrance of a ritual affects the character and atmosphere of a place, even if the ritual is not executed any longer. The relationship of ritual and place seems to be intimate, complex, and somehow reciprocal. Place and ritual affect each other reciprocally. The specific attributes of place and context offer necessary conditions for ritualization, and the ritual practice includes and transforms the atmosphere and identity of a place where it takes place.

Rituals are, as we see, regulating human movements, and as urban space offers complex potential for many kinds of movements that are interlinked in a sublime way, an analysis of ritual practices and their ability to regulate movements in urban space would generate a lot of new insights. Rituals represent a form of social practice that takes place in the city both as visible and as invisible religion.

We can, for example, compare the spiritual pilgrimage, which in its present practices often is connected to and motivated by environmental values, with modern sport, even if sporting events are both similar to and different from rituals. The pilgrim places his or her walk over the movement as transport. Pilgrims experience their walking as a tool for contemplation, for mental and bodily regeneration and for a more intimate integration with nature, which we could easily compare to the modern city marathons. Even if the runner wants to reach the goal, it is the process of running together with many others that gives meaning to the run and the whole modern marathon. Most of the participants are not professional athletes but ordinary people who have spent a lot of time preparing and training. They are sweating, they are limping, they are faltering, but the experience of partaking in a common exertion makes their pain meaningful and even sensual. The marathon appears as a deeply ritualised form of common movement in urban space. There is no doubt that it changes the city itself, and that it transforms the economy of the modern urban space for trading, transport, and symbolic encounters into something else. Like a carnival, a religious festival, or a course for celebrating the winning football team, the urban lived space radically changes its atmosphere when the mass just runs.

7 Rituals in Transit to the Eocene?

Whereto might ritual power, as reflected and circumscribed in this chapter, take us in the future? The author would be happy if the reader could leastwise foreshadow the potentials of a further exploration of the interaction of rituals and aesth/ethics.

Might rituals assist us in offering new modes of executing power-with rather than power-over others? Might they be able to create new places for conversion, creativity and socio-ecologic transformation? How can rituals fabricate meaning in their own sense as well as in a critical and constructive reciprocity with rational explanation? How can rituals, globally and locally, contribute to the urgently needed great transformation into a just, deep-sustainable common future and Earth? Might rituals in synergy with environmental aesth/ethics assist in moving towards a future beyond the Anthropocene, a truly life-enhancing »Eocene« (Bergmann 2020: Chapter 8)?

—Sigurd Bergmann, Norwegian University of Science and Technology, Norway

References

Altner, Günter, *Tod, Ewigkeit und Überleben*, Heidelberg, 1981.

Bell, Catherine M., *Ritual: Perspectives and Dimensions*, New York and Oxford, 1997.

Bergmann, Sigurd, ›Atmospheres of Synergy: Towards an Eco-Theological Aesth/Ethics of Space‹, *Ecotheology: The Journal of Religion, Nature and the Environment*, Vol. 11, No. 3, 2006, pp. 327–57.

Bergmann, Sigurd, *Raum und Geist. Zur Erdung und Beheimatung der Religion: Eine theologische Ästh/Ethik des Raums*, (Research in Contemporary Religion, Vol. 7), Göttingen, 2010.

Bergmann, Sigurd, »Religion in the Built Environment: Aesth/Ethics, Ritual and Memory in Lived Urban Space«, in: ed. Van Herck, Walter and Gomez Popescu, Liliana, *The Sacred in the City*, London and New York, 2012a, pp. 73–95.

Bergmann, Sigurd, »Fetishism Revisited: In the Animistic Lens of Eco-pneumatology,« *Journal of Reformed Theology*, Vol. 6, 2012b, pp. 195–215.

Bergmann, Sigurd, *Religion, Space & the Environment*, London, 2014.

Bergmann, Sigurd, »›Millions of Machines are Already Roaring‹—Fetishized Technology Encountered by the Life-Giving Spirit,« in: ed. Deane-Drummond, Celia, Bergmann, Sigurd and Szerszynski, Bronislaw *Technofutures*,

Nature and the Sacred: Transdisciplinary Perspectives, London and New York, 2015, pp. 115–137.
Bergmann, Sigurd, *Weather, Religion & Climate Change*, London and New York, 2020.
ed. Bergmann, Sigurd and Sager, Tore, *The Ethics of Mobilities: Rethinking Place, Exclusion, Freedom and Environment*, Aldershot, 2008.
Brecht, Bertolt, *Das Leben des Galilei*, 1939–55, Frankfurt am Main, 1998.
ed. Brunn, Stanley D., *The Changing World Religion Map. Sacred Places, Identities, Practices and Politics*, New York, 2015.
Buttimer, Anne, »Nature, Culture and the Quest of the Sacred«, in: ed. Brunn, Stanley D., *The Changing World Religion Map. Sacred Places, Identities, Practices and Politics*, New York, 2015, pp. 71–82.
Daly, Herman E. and Cobb Jr., John B., *For the Common Good: Redirecting the economy toward community, the environment, and a sustainable future*, 2nd ed., Boston, 1994.
Dark Mountain. https://dark-mountain.net/ [2.01.2019]
Faith activism conference. https://sites.google.com/view/faith-activism-conference-2017/home [2.01.2019]
Grimes, Ronald L., »Ritual theory and the environment«, *The Sociological Review*, Vol. 51, Issue Supplement 2, 2003, pp. 31–45.
Grimes, Ronald L., *The Craft of Ritual Studies*, New York, 2014.
Harvey, Graham, *Animism: Respecting the Living World*, New York, 2006.
Hornborg, Alf, »Submitting to Objects: Fetishism, Dissociation, and the Cultural Foundations of Capitalism,« in: ed. Harvey, Graham, *The Handbook of Contemporary Animism*, Stocksfield, 2013, pp. 244–259.
Hornborg, Alf, »Technology as Fetish: Marx, Latour, and the Cultural Foundations of Capitalism«, *Theory, Culture & Society*, Vol. 31, No. 4, 2014, pp. 119–140.
Hornborg, Anne-Christine, *A Landscape of Left-Overs*, (Lund Studies in History of Religions, Vol. 14), Lund, 2001.
Hornborg, Anne-Christine, *Ritualer: Teorier och tillämpning*, Lund, 2005.
Hornborg, Anne-Christine, *Mi'kmaq Landscapes: From Animism to Sacred Ecology*, Aldershot, 2008.
Hornborg, Anne-Christine, »Protecting Earth? Rappaport's Vision of Rituals as Environmental Practices«, *Journal of Human Ecology*, Vol. 24, No. 4, 2008, pp. 275–283.
Hornborg, Anne-Christine, »Selling Nature, Selling Health: The Commodification of Ritual Healing in Late Modern Sweden«, in: ed. Bergmann, Sigurd and Kim, Yong-Bock, *Religion, Ecology & Gender: East-West Perspectives*, (Studies in Religion and the Environment, Vol. 1), Berlin, 2009, pp. 109–129.
Ingold, Tim, *The Perception of the Environment: Essays in Livelihood, Dwelling and Skill*, London and New York, 2000.
Insoll, Timothy, *The Oxford Handbook of the Archaeology of Ritual and Religion*, Oxford, 2011.
Luther, Martin, *The Large Catechism*, 1529, http://bookofconcord.org/lc-1-intro.php [14.01.2019].

Maxwell, Nicholas, *From Knowledge to Wisdom: A Revolution in the Aims and Methods of Science*, Oxford, 1984.
Petersson, Anna, »Things as Links between Person and Place«, in: ed. Bergmann, Sigurd, Scott, Peter M., Jansdotter Samuelsson Maria and Bedford Strohm, Heinrich, *Nature, Space and the Sacred: Transdisciplinary Perspectives*, Aldershot, 2009, pp. 131–144.
Rappaport, Roy A., *Pigs for the Ancestors: Ritual in the Ecology of a New Guinea People*, New Haven, 1968.
Rappaport, Roy A., *Ritual and Religion in the Making of Humanity*, Cambridge, 1999.
Schneidewind, Uwe, *Die Große Transformation: Eine Einführung in die Kunst des gesellschaftlichen Wandels*, Frankfurt am Main, 2018.
Soja, Edward W., *Thirdspace: Journeys to Los Angeles and Other Real-and-Imagined Places*, Malden and Oxford 1996.
Soja, Edward W., *Postmetropolis. Critical Studies of Cities and Regions*, Oxford, 2000.
Tillich, Paul, *Systematic Theology, Three Volumes in One*, Chicago, 1951, 1957, 1963.
Tillich, *Paul, Dynamics of Faith, New York, 1957.*

The Self-with-others and Environmental Ethics

> The land and sea, the animals, fishes, and birds, the sky of heaven and the orbs, the forests, mountains, and rivers, are not small themes [...] but folks expect of the poet to indicate more than the beauty and dignity which always attach to dumb real objects [...] they expect him to indicate the path between reality and their souls.
>
> (Walt Whitman, Preface to *Leaves of Grass*, 1855)

The land and sea, the animals, fishes, birds, and everything else – in other words, all things natural – are definitely not small themes. Nature is not a small theme. Neither is the human being, or the self. But perhaps the poet is not the only one to indicate the path between reality and souls, as philosophers might also have a say in this.

I believe that any discussion about environmental ethics has to start from an examination of the relation between human beings and nature. Does nature have an intrinsic value, independent of human beings? Or is the value of nature the result of the (economic, emotional, etc.) investment made by human beings? These two positions have informed the debate in environmental ethics for a long time, but are they adequate as points of departure? Of course, the position we choose informs the direction of our discourse on environmental ethics, but perhaps it is time for us to take a step back and start afresh with a reconsideration of the connection between nature and human beings. After all, as Inutsuka suggests, »[t]o re-examine our concept of the environment is at the same time to re-examine our concept of humanity« (Inutsuka 2017: 88). To this, I add that to re-examine our concept of the environment also means to re-examine our concept of the self.

In this essay[1], I discuss the notion of *hito* (human being) and the concept of the self as they appear in the work of Japanese philosopher Andō Shōeki 安藤昌益 (1703–1762). One of Shōeki's most original and significant contributions to world philosophy is his understanding of the human being, the notion of *hito* envisaged not only as a hub of man-and-woman fused together, but also as an all-encompassing self that expands into all of humankind. His major work, *Shizen shin'eidō* 自然真営道, proposes a vision of the world where two different realms exist: *shizen no yo* 自然の世 (›the World of *shizen*‹), and *shihōsei* 私法世 (»the World of the Private Law«). *Shizen no yo* is the world of a primordial, pristine nature where all forms and manifestations of energy and life exist in an ideal, uncorrupted state, whereas *shihōsei* represents human society, vitiated by the introduction of man-made, self-serving laws and thus marred by an estrangement from nature. There are, of course, differences between the *hito* in the World of *shizen* and the *hito* in the World of the Private Law—the former is integral to nature, complete in its *isness*, atemporal, non-relative and ahistorical; the latter is divorced from nature, alienated because of the rule of self-serving laws and ideologies, and burdened by the weight of history.

Shizen shin'eidō is a complex work abundant in ideas, concepts and notions that sustain a variety of interpretations. For example, Shōeki can be seen as an advocate of physiocracy, or an »agricultural philosopher«, particularly because of the concept of *chokkō* 直耕 (»straight cultivation«). But he is also an outspoken social critic, dissatisfied with the state of things in Edo period Japan[2], especially with the class system and with the destitution of the farmers, which he criticises by proposing a vision of the world where any social hierarchy is virtually impossible. Last but not least, he is also a naturalist philosopher who puts forth an image of nature as a self-sufficient, complete realm governed exclusively by natural principles and forces.

Here I propose an analysis of Shōeki's vision of the human being *within* the World of *shizen* in an attempt to find hints that might help us better understand the relation between human beings and nature, as well as the rationale for environmental ethics[3]. I develop my argument in two steps. First, I analyse the concept of the human being, concentrating on the principle of *gosei* 互性 (»mutual natures«). I suggest that Shōeki's understanding of the human being is three-leveled, spanning from the single individual to the whole of humankind, and I propose the term *homo naturalis* to refer to this interpretation. Also, I posit that the self of this *homo naturalis* can only be understood as self-with-others. In the second step, I discuss the principles that govern and structure the existence and the conduct of the human being in relationship with *shizen*, paying special attention to the notion of *chokkō* 直耕 (»straight cultivation«). My conclusion is that re-examining the concept of the self and redefining it as self-with-others can also shed new light on the issue of responsibility toward the environment; for Shōeki, *hito* is an integral part of *shizen*, and therefore any discussion about ethics must be re-framed to include the human being. I also suggest that the ›anthropocentric/non-anthropocentric‹ dichotomy should be overcome, as the basis for any kind of ethical decision or moral judgment lies *within the human being as part of nature.*

1 The Terminology of Nature

The title of Shōeki's work *Shizen shin'eidō* has been translated in various ways: *The Way of the Operation of the Self-acting Truth* (Yasunaga 1992), *Grand Traité du Shizenshinei'dō* (Joly 1996), *The Way of Natural Spontaneity and Living Truth* (Heisig 2011), *The Way of the Five Processes and Unitary Generative Force Advancing and Retreating* (Tucker 2013) etc. What all these translations have in common is that none of them uses the word ›nature‹ to render the Japanese *shizen*, and perhaps we should ask ourselves why.

In contemporary Japanese, *shizen* 自然 is the word generally used to render the English ›nature‹, and, morphologically, it functions just like ›nature‹, i.e. as a noun. However, as Yanabu (1977) points out, the understanding of the term *shizen* in Edo period Japan was not ›nature‹, but rather ›spontaneously acting/doing‹, and it was generally used as an adjective, or an adverb. In order to designate what we now understand by ›nature‹, the Japanese made use of various other terms, many of them of Chinese origin and most of them with a philosophical tinge (Daoist, Confucianist, or Buddhist): *tenchi* 天地 (›Heaven-and-Earth‹), *banbutsu* 万物 (›the myriad things‹), *sansen-sōmoku* 山川草木 (›mountains, rivers, plants and trees‹), *senga-daiji* 山河大地 (›mountains, rivers, and the Earth‹), *zōbutsu* 造物 (›things that are made/created‹)[4] etc. The same compound 自然 was actually read *jinen* at the time, and, as Tellenbach and Kimura (1989: 157) show, it had a strong Buddhist connotation and it referred rather to 1) that which exists ›without human intervention‹, and 2) that which exists ›of/from oneself/itself so, truly so‹. Shōeki, however, forces us to read it as *shizen:* in a text titled *Kakuryū sensei inkeisho (Master Kakuryū's Classic)*[5], included in volume 16 of Shōeki's complete works as part of the *Hachinohe shiryō (Hachinohe documents)*, he states that it should be read *shizen:* 「父字ノ『シ』ト、母字ノ『ゼン』」(ASZ 16: 302). ›The first ideograph is [read] *shi*, the second *zen*‹.

So, does *shizen/jinen* actually mean ›nature‹ ? Well, yes and no. Yes, if we absolutely want to find an English equivalent—it would probably be the term closest to our understanding of ›nature‹. And no, because it does not designate the same notion. As Tellenbach and Kimura put it: ›Contrary to the object-oriented meaning of nature, *Shi-zen/Ji-nen* has no such meaning. It never signifies the object as such, but presents only its respective manner of being and becoming‹ (Tellenbach and Kimura 1989: 157). In other words, *shizen/jinen* rather refers to the way in which the world *presents itself* to us, not to its material, substantial component. And this *presenting* implies that nature is not a fixed entity, that it should be viewed more like a dynamic process through which everything that surrounds us is constantly *spontaneously being and becoming*. This is, most likely, the strongest reason why all the translations of the title *Shizen shin'eidō* avoid the word »nature« and instead propose alternatives, which are all justified by Shōeki's understanding of *shizen*[6].

In this essay, I will take the same position and refrain from using ›nature‹ when discussing Shōeki's philosophy. Instead, I will use the original *shizen*, especially when describing the notion of *hito* and the concept of self, since they appear as such precisely because they are realised within *shizen*. I ground my choice in Tremblay's suggestion that, when translating Japanese philosophy—or any other philosophy for that matter, I might add—we should force and expand the limits of language, as not all concepts and notions have (nor should they have) perfect equivalents (Tremblay 2008: 242):

> À la façon des philosophes de l'ère Meiji qui transformèrent leur propre langue (création de néologismes et de distinctions à partir des caractères chinois, nouvelles significations données aux vieux vocables, altération de la syntaxe, bref, extension des limites de la langue), les traducteurs actuels de la philosophie japonaise doivent accomplir le même type de travail au niveau de leurs langues maternelles respectives [...].

But what kind of vision of the world does Shōeki put forth, after all? And how does the human being fit in this world? In his own words, *shizen* is a realm made up of spontaneous energies that circulate ceaselessly back and forth from Heaven to Earth, passing, in between, through the ground and the seas in the middle and begetting all creatures (ASZ 1997: I: 63–64):

Shizen is the special name of the Subtle Way of mutual natures. But what are the mutual natures? They are the spontaneous movement of the primary matter of earth *(dokasshin)*—which is beginningless and endless—which advances and retreats to a greater or lesser degree. [The primary matter] thus creates the four elements: when it advances a little: wood; when it advances a lot: fire; when it retreats a little: metal; when it retreats a lot: water. Spontaneously, by advancing and retreating, [the four elements] create the eight energies, which are mutual natures. [...] This [dynamic process] is the Subtle Way. It is ›subtle‹ because of the existence of the mutual natures, and it is a ›Way‹ because of the interaction of these mutual natures. This is the spontaneous movement of the primary matter of earth—which cannot be taught or learned, which does not increase or decrease—which is created by itself. Therefore, this is called *shizen*[7].

One of the most important terms in this fragment—apart from *shizen*—is *kasshin* 活真 (which can also be read *ikite makoto*). *Kasshin* is the term coined by Shōeki to designate the primary matter, the fabric of all existence. *Katsu/ikite* 活 represents vitality and dynamism, whereas *shin/makoto* 真 represents the substantiality and materiality of existence. The way in which *kasshin* advances and retreats is regulated by *gosei*, the »mutual natures«, which designates the fundamental way in which *shizen* functions and plays an important part in the ontology of the *hito* and of the self.

2 The Self-with-others

Gosei consists of *go*, which means ›reciprocal‹ or ›mutual‹, and *sei*, which means ›embedded feature‹, ›inner characteristic‹, ›immanent disposition‹, ›interior(ised) reciprocity‹ etc. For Shōeki, this notion implies that absolutely all constitutive elements found in *shizen* are in a relationship of functional reciprocity, from Heaven and Earth to the advancing and retreating energies, to fire and water, to man and woman. Therefore, the only connection that can obtain between any two entities present in *shizen* is one of ›mutual natures‹, which means that each of the two entities contains within itself the essence of the other. Thus, they are neither distinct, nor identical; they exist as two sides of the same coin, separated yet inseparable.

Perhaps the best example of ›mutual natures‹ is Shōeki's understanding of the human being, of man and woman as a single self of man-and-woman (ASZ 1997: I: 113–114):

> The primary matter is constantly acting in a subtle manner through the mutual natures of advance and retreat, without a moment's pause. [...] The nature of the man is the woman, and the nature of the woman is the man—with their mutual natures of man and woman, they are the human being as manifestation of the primary matter.

As we can see from this fragment, for Shōeki the human being—i. e. man-and-woman-as-a-single-person—is not merely a temporary pairing of two different entities. He writes the concept with two ideographs (*otoko* and *onna*, 男女), but he specifies that this compound should not be read *danjo*, but *hito*. Thus, *hito* is more than the sum total of man plus woman as two distinct elements, because it represents in fact a fusion of two forms of existence that, while separated and heterogeneous, contain within themselves the valency needed to be combined with each other in an indissoluble union. Just as Heaven and Earth, water and fire, or the flows of energies are inextricably linked in pairs as ›mutual natures‹, so are man and woman merged into one, in accordance with *gosei* as an ontological principle that underpins all of existence. As *otoko* is embedded within *onna* and *onna* is embedded within *otoko*, the *hito* resulted from the amalgamation of the two is the epitome of ›mutual natures‹ and the most easily recognizable manifestation of this notion.

Since man and woman both contain the fundamental inner characteristics of the other—which thus become embedded features of their very own essence—they retain the valency and potentiality for union but at the same time conserve and perpetuate their own, separate identity. *Gosei* is therefore a *principle of mutual independence* in which each of the two entities supports and enhances all the features, characteristics and qualities of the other, thus underlying the image of a *homo naturalis* reconnected with *shizen* and reinstated as a full-fledged component of its realm.

Since *hito* is a manifestation of ›mutual natures‹ and, as such, part of the intricate system of *shizen*, it is clear that the world envisaged by Shōeki can never be an anthropocentric universe. The self is not a *res cogitans*, a sentient being contemplating the world (or the *res extensa*, as Descartes would put it), but actually a *homo naturalis*, a mere constituent of this world placed on an equal footing with all the other elements, from plants to crawling creatures, from fishes to rivers. Moreover, not only are one man and one woman fused together into one single person, but all human beings are in a relationship of ›mutual natures‹. The only connection that a *hito* can establish with another *hito* is one of mutual independence, and thus the concept acquires a whole new dimension as it is used to refer to society at large, not to just one single unit. The notion of the human being understood as *hito* does refer to the ontological characteristics of the individual, but at the same time it encompasses all of humankind, viewed as a complex web of interwoven reciprocities. Therefore, all *hito* are one *hito*, precisely because they are interlinked by the principle of *gosei*—ontologically and epistemologically, the human being can not just *exist without* the other human beings. One *hito* exists in its isness solely because all the other *hito* are at the same time in a relationship of mutuality with it.

Since *shizen* itself is beginningless and endless, the question of time is irrelevant and, as a consequence, the notion of history is meaningless. Therefore, the human being (and any other form of existence) is ahistorical and non-relative, and there can be no value judgment with regard to its existence and presence in the world, which further means that any type of hierarchy within the realm of *shizen* is fundamentally impossible and inapposite. Shōeki illustrates this idea by using the term *nibetsu naki* 二別無キ (›no distinction‹) to emphasise the fact that there is no difference between man and woman (or between superior and inferior). *Nibetsu naki* implies that there is no divergence between the two entities fused together as ›mutual natures‹. It suggests, at the same time, that both man and woman, *hito* and *hito*, *hito* and the others exist as individual, distinct entities as well. They are simultaneously homogeneous and heterogeneous, innate and immanent within each other. This entails another feature of the human being in Shōeki's vision: because of the principle of *gosei*, the essence of the *hito* is disseminated within all the others, and therefore the self of one human being is at the same time the self of all other human beings. This means that the whole of humankind, while dispersed into a multitude of individual, separate manifestations, is in fact one single person, a universal, global ›I‹, an all-encompassing self that epitomises human nature. This is the *homo naturalis*, intrinsic to *shizen* as an integral component, unfettered by hierarchies or value systems, rooted at the same time within the self and the others, and affirmed and defined through the unmediated interaction with the others in accordance with the principle of ›mutual natures‹.

How is this vision of *hito* relevant for environmental ethics? There are, of course, numerous approaches to the question of the locus and role of the self in disciplines as varied as philosophy, psychology, psychiatry, medicine, neurobiology etc. One of the theories I would like to bring into the discussion here is the »relational self« in psychology as put forth by Andersen and Chen (2002). Drawing from previous work on interpersonal approaches to the self and grounding their thinking in social-cognitive views of personality, the authors propose a view of the self which is closely linked with the phenomenon of transference (Andersen and Chen 2002: 638):

> The theory posits that individuals possess multiple selves in relation to the various significant others in their lives, each linked in memory with a particular significant other. It is because linkages exist between the self and significant-other representations that when a significant-other representation is activated in an encounter with a new person, associated self-with-significant-other knowledge is set into motion. As a result, the relevant relational self comes into play, as manifested in the individual's emotional, motivational, and behavioral responses in the interpersonal encounter. Different relational selves thus unfold dynamically across interpersonal contexts.

As psychologists, for Andersen and Chen the focus is on the individual's interpersonal patterns and, therefore, on his or her personality. They propose the existence of a ›repertoire of relational selves‹ and suggest that each individual invokes and relies on a particular self from this repertoire when dealing with different significant others in different social contexts and situations. I think that this theory of the ›relational self‹—with two essential amendments—can be useful in trying to better understand Shōeki's notion of *hito* and its relation with *shizen*.

The first amendment I propose is that for Shōeki the relation is not limited to the ›significant other‹. As I have shown, *gosei* is the only mode of existence within *shizen*, the only kind of relationship that can obtain between its various constituting elements. Therefore, one particular, individual self cannot exist *as such*, firstly because it already represents a fusion between man and woman, and secondly because in its broadest instantiation it includes all of humankind. Supposing *ad absurdum* that such a self exists, then it cannot be complete in itself, as it is ontologically interlinked with all the other selves. In other words, the self for Shōeki is necessarily a *self-with-others*.

The second amendment is that, if we adopt Shōeki's idea of *gosei*, then there can be no such thing as a ›repertoire of relational selves‹. The *hito* in the realm of *shizen* does not choose between different selves, simply because there is nothing to choose from. The entity of a man-and-woman-as-a-single-person does not construct its personality by navigating and negotiating between different selves according to the social situation, as it is intrinsically and inextricably linked with all the others: if your self can only exist as a *self-with-others*, then there is nothing to negotiate in this kind of relationship.

To explore the link between this notion of self-with-others and environmental ethics, we need to look into the locus of the agency of the *hito* that Shōeki proposes. And in order to do that, we should take a closer look at *chokkō*, a concept which informs Shōeki's understanding of the relationship between human beings and *shizen*. With the literal meaning of ›straight/direct cultivation‹, *chokkō* is a concept coined by Shōeki to designate all activities of labor or production, from tilling the land to harvesting crops. However, *chokkō* is also used in a broader sense to refer comprehensively to the sum of all creative activities, whether it be the creative energies and ontological capacities of *shizen*, or the tasks and actions which underpin the existence of all forms of life, from human beings to plants.

In a universe governed by *chokkō*, any human act represents a reproduction and a continuation of the ontological movements of *shizen* and, thus, *chokkō* refers comprehensively to the activities and power of creation of both Heaven and human beings. The first and most basic meaning of ›straight cultivation‹ is undoubtedly the agricultural one, but this meaning becomes subaltern through the abstractisation of the term. Thus, when Shōeki states that ›straight cultivation‹ is the only way for human beings as constitutive elements of *shizen*, he is talking not only about literally, physically tilling the land, but also about something more comprehensive, subtle and intangible, a kind of vague, yet pervasive awareness of all forms of existence that their *raison d'être* is to be in accord with *shizen*.

Here is how Shōeki himself defines *chokkō* (ASZ 1997: I: 64–65):

[The energies] ascend to Heaven, and after ascending they descend and, in accordance with the land in the middle, they acquire the three directions—descending *(tsū)*, lateral, and ascending—and create and produce grains, human beings *(hito)*, the four types of creatures, and vegetation. This is the creative power *(chokkō)* of the primary matter, beginningless and endless. Consequently, Heaven-and-Earth, the stars, the planets, the sun and the moon—in other words, the Heaven-and Earth which moves in accordance with the three directions—are all manifestations of the energy of the primary matter.

Shōeki's discontent with the world he was living in determined him to put forth the concept of ›straight cultivation‹ as a means of making sense of society's ills while at the same time imagining a primeval, prelapsarian World of *shizen* in an endeavor to reestablish the connections among its various components. Since *chokkō* is the only true way in which the *homo naturalis* can live, it also signifies a rehabilitation of the intricate web of relationships between human beings and *shizen*, and a reinstatement of the human being *qua* human being. To accomplish this, Shōeki invests ›straight cultivation‹ with meanings and nuances that exceed and transcend the simple notion of tilling the land; *chokkō* thus becomes all-pervasive and ubiquitous, and its ontological capacity and creative powers become embedded within every gesture or action performed by the human being, including apparently mundane tasks such as cooking (ASZ 1997: I: 81–82): ›The only purpose of the subtle action that takes place in the hearth is to prepare the grain to eat. It is splendidly clear that this is [an example of] the straight cultivation of the primary matter of earth in the hearth‹.

To sum up, the *hito* in Shōeki's vision of the world—the *homo naturalis*—is a notion that comprises three different stages: first of all, it is the fusion between man and woman as distinct, yet inseparable entities; secondly, it is the interaction between the single person resulted from the fusion and all the others; and thirdly, it is the all-encompassing single self resulted from this interaction. In all these instances, the existence of the human being is governed by the principle of ›mutual natures‹, while its relationship with the realm of *shizen* is regulated by the principle of ›straight cultivation‹.

But where is the locus of the sense of agency for the self-with-others? Is *hito* an agent in its relation with *shizen?* When discussing different theories of the notion of ›agent‹, Benson notices that they ›all presuppose that agents are necessarily identified with or present in some privileged volitional states or complex of states‹ (Benson 1994: 654–655). In a more recent study Gallagher notes that, in general, there can be identified three meanings of agency: first, the sense of agency as a first-order experience linked to intentional aspect (task, goal, etc.); second, as first-order experience linked to bodily movement; and third, as a second-order, reflexive attribution. (Gallagher 2007: 354–355) The author then goes on to suggest that the sense of agency is not, in fact, reducible to somatic processes, nor is it simply the product of higher-order processes as in Descartes' dualism or in Kant's transcendental ›I‹, and proposes instead a hybrid understanding of the sense of agency which consists of a complex combination of outward-oriented signals, inward-oriented feedback, and intentional feedback (Gallagher 2007: 354): »efferent signals, sensory (afferent) feedback, and intentional (perceptual) feedback. If any of these contributory elements fail, or fail to be properly integrated, then we can get a disruption or disturbance in the sense of agency«.

However, if we agree with Shōeki that for the self-with-others the only way to exist within *shizen* is *chokkō*, then there can be no question of a ›volitional state‹ or ›intentional feedback‹ since *chokkō* regulates every aspect of the presencing of the *hito*—the issue of volition or intention becomes completely irrelevant in this context. For the same reason, there can be no hybrid sense of agency, as all elements that might contribute to it are in fact subsumed to *shizen* through *chokkō*, from bodily movements to higher-order processes. The *hito* as self-with-others breathes, moves, makes tea, reasons—in a word, *exists*—within the realm of *shizen* only through ›straight cultivation‹. As such, it is a self without agency in any understanding of the term. Or, to put it differently, it is a self that has relinquished its agency and entrusted it to *shizen*.

3 *Gosei*, *chokkō* and Environmental Ethics

As we have seen, the concept of *hito* lies at the very core of Shōeki's philosophy. It is a complex notion describing the human being on three different levels, closely interlinked with other notions such as ›mutual natures‹, or ›straight cultivation‹, and integrated into the vast fabric of *shizen*. This *homo naturalis* is in a relation of mutual independence with all the other human beings, forming together a self-with-others. Moreover, if—as Shōeki suggests—we expand the scope of *gosei* beyond the realm of the human, then we can even talk about a self-with-everything within the world of *shizen*. Since ›mutual natures‹ is the only kind of relation that can obtain between any two entities, then the universal self of humankind is extended to the non-human as well, and the self-with-others includes the land and sea, the animals, fishes, and birds, the sky of heaven and the orbs, the forests, mountains, and rivers …

If the basis for the presencing of the self lies in *shizen;* if the self cannot be anything else but a self-with-others; if the agency of the self is entrusted to *chokkō;* and if, thus, the self-with-others becomes the self-with-everything, then the issue of responsibility toward *shizen* must be reconsidered starting from these premises. If the human being's *raison d'être* is to be within *shizen* as an integral, inseparable part of it, then any action that a human being might take in whatever context is an action directed at the same time at *shizen* and at the self itself. When I am confronted with the decision whether I should fell a tree or not, I need to consider the fact that I am, at least partially, also deciding whether to fell myself. When I kill a bird and eat it, I also kill at least a part of myself. When I hurt *shizen* in the smallest way, I also hurt myself. But I should never get that far: my understanding that my agency lies within ›straight cultivation‹ should prevent me from even contemplating such choices.

If Shōeki is correct, then this means two things for environmental ethics: first of all, we must adjust our definition of ›environment‹ to include human beings *within* it, not outside of it. *Hito* is not a *res cogitans* observing nature from an external realm, but an inherent part of it. To go back to the question I started with: the value of nature is neither the result of the investment made by human beings, nor is it an intrinsic value independent of human beings—both these positions presuppose a human being separated from nature. The value of nature, its beauty and dignity, is intrinsic *because* human beings are a constitutive part of *shizen*—not independent from it, not investing anything in it, just existing or, to be more exact, just *presencing* themselves in it. And this leads us to the second thing: in considering our ethical positioning toward the environment, we must start from the premise that we *are* the environment. More precisely, we are ourselves because *shizen* is the ontological and epistemological principle that underpins our existence and therefore we are part of it just as much as it is part of us. In this sense, the discussion whether ›nature‹ is ›anthropocentric‹ or ›non-anthropocentric‹ is superfluous and futile, as there is no ›centre‹ and no ›periphery‹. If the self-with-others is the self-with-everything, then the self is everything and everywhere. The ethics of the environment is the ethics of human beings, and vice versa.

There is substantial potential for research in the field of environmental ethics in general and in Japan in particular. But I believe that this kind of research should be accompanied by a process of re-examination of fundamental concepts such as ›self‹, ›nature/*shizen*‹, and the relation between them. Transforming current concepts, adding new meanings to old notions and forcing the limits of language with new translations are tasks we should continue to engage in, constantly. This will open new doors into our understanding of the environment and of its ethics.

—Roman Pașca, Graduate School of Letters, Kyoto University, Japan

References

ASZ *Andō Shōeki Zenshū* [›Complete works of Andō Shōeki‹], Tokyo, 1997.
Adeney Thomas, Julia, *Reconfiguring Modernity: Concepts of Nature in Japanese Political Ideology*, Berkeley, 2001.
Andersen Susan M. and Chen, Serena, »The Relational Self: An Interpersonal Social-Cognitive Theory«, *Psychological Review*, Vol. 109, No. 4, 2002, pp. 619–645.
Benson, Paul, »Free Agency and Self-Worth«, *The Journal of Philosophy*, Vol. 91, No. 12, 1994, pp. 650–668.
Gallagher, Shaun, »The Natural Philosophy of Agency«, *Philosophy Compass* 2/2, 2007, pp. 347–357.
ed. Heisig, Jim, Kasulis Tom and Maraldo, John, *Japanese Philosophy. A Sourcebook*, Honolulu, 2011.
Inutsuka, Yu, »Sensation, Betweenness, Rhythms. Watsuji's Environmental Philosophy and Ethics in Conversation with Heidegger«, in: ed. J. Baird Callicott and McRae, James, *Japanese Environmental Philosophy*, Oxford, 2017, pp. 87–104.
Joly, Jacques, *Le naturel selon Andō Shōeki. Un type de discours sur la nature et la spontanéité par un maître-confucéen de l'époque Tokugawa, Andō Shōeki, 1703–1762*, Paris, 1996.
Kawai, Hayao, *Dreams, Myths and Fairy Tales in Japan*, Einsiedeln, 1995.
Morris-Suzuki, Tessa, *Reinventing Japan: Time, Space, Nation*, New York, 1998.
Pașca, Roman, »*Homo Naturalis*—Andō Shōeki's Understanding of the Human Being«, in: ed. Morisato, Takeshi, *Critical Perspectives on Japanese Philosophy*, Nagoya, 2016, pp. 78–99.
Tellenbach Hubertus and Kimura, Bin, »The Japanese Concept of ›Nature‹«, in: ed. J. Baird Callicott and Ames, Roger T., *Nature in Asian Traditions of Thought*, New York, 1989, pp. 153–162.

Spinoza, Baruch, *Ethics, Part I*, Prop. 29, Scholium, trans. by Edwin Curley, London, 1996.
Tremblay, Jacynthe, »Liminaire«, *Laval théologique et philosophique* Vol. 64, No. 2, 2008, pp. 237–243.
Tucker, John A., »Andō Shōeki's Agrarian Utopianism: An East Asian Philosophical Contextualization«, *Taiwan Journal of East Asian Studies*, Vol. 10, No. 1 (Issue 19), 2013, pp. 53–86.
Yanabu, Akira, *Hon'yaku no shisō – ›shizen‹ to ›nature‹* [›The philosophy of translation—»shizen« and »nature«‹], Tokyo, 1977.
Yasunaga, Toshinobu, *Andō Shōeki – Social and Ecological Philosopher of Eighteenth-Century Japan*, New York: Weatherhill, 1992.

Notes

[1] This work was supported by JSPS KAKENHI Grant Number 17854234.
[2] The Edo (or Tokugawa) period in Japanese history starts in 1603 when the Tokugawa shoguns become *de facto* rulers of the country, and ends in 1868 with the Meiji Restoration. It is characterised by the *sakoku* (›closed country‹) policy of isolationism imposed by the shogunate, but also by demographic and economic growth, a relatively strict social hierarchy, and political stability. The intellectual landscape is dominated by Confucianism (in particular the Neo-confucianist school of *Shushigaku*), but it also sees the development of disciplines such as *kokugaku* (›national learning‹) and *rangaku* (›Dutch learning‹.)
[3] I have discussed some of these themes before (Paşca 2016), but my focus was the relation between the *homo naturalis* and the three flows of energy: descending, lateral, and ascending *(tsūki, ōki,* and *gyakki).*
[4] For a more detailed discussion of *jinen/shizen*, cf. also Adeney Thomas (2001); Joly (1996: 77–156); Kawai (1995); Morris-Suzuki (1998: 40); Yasunaga (1992: 350–352).
[5] Kakuryūdō Ryōchū 確龍堂良中 was one of the pseudonyms Shōeki used.
[6] This description of nature may sound somewhat similar to Spinoza's notion of *natura naturans*, but they are in fact fundamentally different, in that for Spinoza nature is ultimately God, whereas for Shōeki there is no such thing as a transcendental entity of any kind. Spinoza tells us that *natura naturans* is ›what is in itself and is conceived through itself, or such attributes of substance as express an eternal and infinite essence, that is […] God, insofar as he is considered as a free cause‹, and then opposes it to—and at the same time, pairs it with—*natura naturata*, which is ›whatever follows from the necessity of God's nature, or from God's attributes, that is, all the modes of God's attributes insofar as they are considered as things which are in God, and can neither be nor be conceived without God‹. In Shōeki's vision, however, there is no *naturata* dimension to nature, as *shizen* is complete in itself and not the equivalent of some divine creation.
[7] All translations from *Shizen shin'eidō* are from Yasunaga (1992), modified and adapted to better fit the original. Unless otherwise indicated, all other translations are mine.

Climate Apocalypticism and the Temporal Sublime

The prospect of climate collapse haunts contemporary culture and political debate today in a way that no environmental threat has before. Such concerns over what the future will bring are well-founded, given the global reach and profound impact of climate change's anticipated effects on human and natural systems, the political and economic challenges of an effective international response, and the escalating risks of deferred action. It is unsurprising, then, that our narratives for thinking about climate change have adopted an apocalyptic structure, by which they continue a defining trend of modern environmental rhetoric. Without implying any scepticism concerning the very real dangers that climate collapse represents, my aim here is to raise some critical concerns over environmentalism's endorsement of apocalyptic rhetoric, taking popular climate apocalypticism as the guiding example. More precisely, my concern is with the temporal structure of the climate collapse narrative and the conception of the sublime that it implies. It is a commonplace to call attention to the linear temporality and tragic telos of apocalypticism in general, and climate change disaster narratives in particular, but the stakes of this temporal framing have received relatively scant attention.[1] To clarify these stakes, I borrow political theorist Ben Jones's schema for ›cataclysmic apocalyptic thought‹, which highlights the role that crisis plays as transition from the corrupt state of the present to an ideal future. I draw from this the implication that climate apocalypticism, and secular apocalypticism more generally, aims to escape from the time of the present, with the weight of its specific history and the future horizon of its projected fears, into a new age and a new time, one symbolised by the counterfactual time of fiction and the incommensurable times of the deep geological past or indefinite future. I see this leap into another temporality as the culminating moment of a homogenisation of time that Jean-Luc Nancy has ascribed to ecotechnical globalization and that he characterises as a ›catastrophe‹ of ›general equivalence‹ (Nancy 2015: 6). Drawing on Nancy's call to think the ›ever-renewed present‹ alongside Kyle Powys Whyte's proposal of ›time spiraling‹ as a living dialogue with our ancestors and descendants, I suggest an alternative figure of temporal justice that refuses the Judgment Day of apocalyptic narratives.

To understand the role that I am granting to the sublime in interpreting the temporal structure of climate narratives, it is first helpful to consider the historical context of our changing relationship with deep time. The fact that long expanses of time confront the human mind with a sublime dimension was recognised by both Hume and Kant, although neither devotes much attention to this experience.[2] Kant's entire treatment of this topic in his pre-critical *Observations on the Feeling of the Beautiful and the Sublime* (2007 [1764]), for instance, appears in the following few lines:

> A long duration is sublime. If it is of time past, it is noble; if it is projected forth into an unforeseeable future, then there is something terrifying in it. An edifice from the most distant antiquity is worthy of honor. Haller's description of the future eternity inspires a mild horror, and of the past, a transfixed admiration (Kant 2007: 26).

In the terms of the *Critique of Judgment* (1987[1790]), this suggests that the past confronts us with an experience of the mathematical sublime, and indeed Kant refers there to past time as an infinite magnitude (Kant 1987: 254), although this later text offers no further mention of time's sublime character. The unforeseeable future, on the other hand, although never mentioned in the *Critique of Judgment,* would be a species of the dynamically sublime, arousing fear in us in a way that is somehow parallel to the elemental examples that Kant favours: threatening rocks, thunderclouds, volcanoes, hurricanes, and the like (Kant 1987: 261).[3] The reference here to Albrecht von Haller's »Uncompleted Poem on Eternity« (2002) suggests that, for Kant, the future is not to be thought as an infinite magnitude since it is progressing towards its end. And, indeed, he returns to Haller in his 1794 text, »The End of All Things«—a rebuke of Prussian millenarian politics—where what is at stake is not a future proceeding to infinity but precisely eternity as the horrifying abyss that opens beyond the edge of time, beyond the Judgment Day that brings the sensible world to its conclusion. Eternity beyond time is unthinkable, and its »frighteningly sublime« character is due in part to its obscurity; yet according to Kant »in the end it must also be woven in a wondrous way into universal human reason, because it is encountered among all reasoning peoples at all times, clothed in one way or another« (Kant 1996: 221). The caution of Kant's tale is to remember that the religious and cultural imagery with which we clothe this notion of eternity must be understood according to the moral order and not in literal or physical terms.

For both Hume and Kant, the sublime past is revealed only through cultural antiquities, never through natural or elemental phenomena. But in the thirty years that separate these sparse references to the temporal sublime in Kant, developments in what would come to be known as geological science were setting the stage for a dramatic reorientation in our relationship with long durations of time. James Hutton's 1788 *Theory of the Earth* famously proposed a concept of geological time with »no vestige of a beginning,—no prospect of an end« (Hutton 1788: 304), and through the writings of his friend and populariser, John Playfair, this newly opened horizon of what would come to be known as ›deep‹ time was characterised from the outset in sublime terms.[4] This discovery of the deep past simultaneously opens the horizons of the far future and our contemporary cultural obsession with apocalypse.[5] Georges Cuvier's evidence for prehistoric extinctions laid the groundwork for Mary Shelley's exploration of human extinction in *The Last Man* (1826), generally recognised as the first secular apocalyptic novel. The genre of apocalyptic speculative fiction inaugurated by Shelley first gained popularity by imagining our demise from natural causes, but the First World War shifted our fantasies toward the prospect of self-annihilation by weapons of mass destruction. And Rachel Carson's *Silent Spring*, written during the lead up to the Cuban Missile Crisis, played a key role in transferring our nuclear anxieties to the emerging threat of ecological collapse.

Contemporary climate apocalypticism is therefore simply the latest phase in our cultural efforts to manage the sublime dimensions of the unforeseeable future. Just as the threat of total nuclear war—what Derrida in 1984 termed the »phantasm of a remainderless destruction« (Derrida 2007: 396)—framed human reality during the Cold War period, so the phantasm of future climate collapse constructs our present today. Ongoing debates over whether to name our contemporary geological period the ›Anthropocene‹ are symptomatic of this transfigured temporal perspective, which offers a vantage from which humanity can hold itself responsible—for the first time—for our long-term ecological transformations of the globe, while raising—also for the first time—the question of our ethical obligations toward an unimaginably distant future. But insofar as contemporary environmentalism relies on an apocalyptic construction of time that represses the temporal sublime, as we will see below, it cannot do justice to these ethical intuitions. At stake is precisely the temporal framing of apocalypticism, which I investigate here through the example of climate apocalypticism in particular.

1 Climate Apocalypticism

During the final week of the 2009 United Nations Climate Change Conference, four Greenpeace activists paraded on horseback through the streets of Copenhagen dressed in costumes representing famine, pestilence, war, and death. Invoking the four horsemen of the apocalypse from the biblical Book of Revelations, their intent was to dramatise the stakes of climate change negotiations. In a press release from Greenpeace International, Sini Harkka of Greenpeace Nordic explained that »The spectre of the four horseman is looming over these climate negotiations. […] Yet world leaders are still failing to grasp the urgency of the crisis« (Greenpeace International 2009). This is but one dramatic example of the widespread use of apocalyptic rhetoric to describe climate change—by activists, the media, scientists, political actors, advertisers, and popular culture. Over the last decade, such rhetoric has been the subject of interdisciplinary scholarly debate, which has focused primarily on whether framing the narrative of climate change in apocalyptic terms helps or hinders efforts to mobilise individual and collective responses. This debate has generated a proliferation of ways of defining the key elements of apocalyptic narrative and a range of interpretations concerning how the climate change variation extends or remakes the earlier apocalyptic narratives that have framed U.S. environmental discourse since the 1960s (e.g., Rachel Carson's *Silent Spring*, Paul Ehrlich's *The Population Bomb*, and so on). It has also drawn new attention to environmentalism's relationship with the nuclear apocalypticism that preceded and engendered it and the long history of religious apocalypticism in Judaism and Christianity.[6]

My interest here is not with the psychological or political efficacy of apocalyptic rhetoric, but with the temporality that it enacts. Nevertheless, the existing debates shed some light on this, especially as they begin to set climate apocalypticism into its broader historical context and to parse out its distinct variants. These debates tend to start from an understanding of ›apocalypse‹ as straightforwardly synonymous with catastrophe, with the end of the world ›as we know it‹, whether that means the end of ›our‹ current standard of living, or the end of human civilization in any historically recognizable form, or the literal extinction of the human species, and so on. And when apocalypse is read as synonymous with catastrophe, the rhetorical deployment of the narrative is understood to be in the service of galvanising individual action and political will through fear and horror at the likely consequences of inaction. This rhetorical strategy can then be criticised as ineffectual or counter-productive fear-mongering along the lines familiar from Nordhaus and Shellenberger (2007).

Representative of this approach is geographer Mike Hulme's frequently cited typology of four Biblically named climate change ›myths‹: Lamenting Eden, Constructing Babel, Celebrating Jubilee, and Presaging Apocalypse (Hulme 2010). While *Lamenting Eden* voices our misplaced nostalgia for pristine nature's lost autonomy, *Constructing Babel* expresses our hubristic desire for increased technoscientific control, most implicit in geoengineering schemes. *Celebrating Jubilee* serves our ›instinct‹ to do the right thing by taking the climate crisis as an opportunity to establish social and environmental justice. And finally, *Presaging Apocalypse* »lends a sense of danger, fear and urgency« to climate discourse by exhorting us to »save the planet« before it is too late (Hulme 2010: 44). But the »counter-intuitive outcome« of this framing, Hulme argues, is »disempowerment, apathy and scepticism among its intended audience« (Hulme 2010: 45) Now, as Hulme is aware, these four narratives rarely operate in isolation, since the apocalyptic mode is driven by a sense of nature's fragility borrowed from *Lamenting Eden*, and the crisis mentality that it engenders is precisely what drives *Constructing Babel* and *Celebrating Jubilee*. So, going beyond Hulme, we might conclude that *Presaging Apocalypse* is not merely one alongside the other myths, but rather the heart of our cultural response to climate change. This would follow from the recognition that environmentalism, at least in the United States, has always defined itself by apocalyptic narratives, and that climate change lends itself to appropriation as *the* paradigmatic apocalypse.[7]

There are several interesting points to note about Hulme's treatment of the *Presaging Apocalypse* ›myth‹. First, he intentionally limits his use of ›apocalypse‹ to its »popular sense«, which he understands as »impending large-scale disaster or destruction« in contrast with the concept's original Greek—and Biblical—usage, meaning simply »disclosure or revelation« (Hulme 2010: 55n7). With this decision, Hulme participates in the trend noted by Stefan Skrimshire of stripping references to apocalypse of their »theological nuances« in favour of their »sensationalist elements«, and particularly of treating such discourses as reducible to fear of the future. What is whitewashed here, as Skrimshire reminds us, is precisely the »complex dramatic structure« of the religious apocalyptic narrative, which includes »the creation of tension between the corruption that is endured in the present age and the hope in the new age that is yet to come« (Skrimshire 2014: 237). The temporal, eschatological element of apocalyptic thinking is precisely to be found in this productive tension ignored by Hulme, and to which we will return.

Furthermore, like others who evaluate the apocalyptic narrative in terms of its rhetorical efficacy, Hulme seems to suggest that such narrative framings are something that we can consciously pick and choose according to our political aims, rather than—at least in part—historical structures that frame our very experience. Obviously, there are those activists, reporters, politicians, and authors who seize on an apocalyptic description as a tactic; but their tactics resonate precisely because of the ways we have come to experience the world through an essentially apocalyptic mode. That this mode is more complex than simple fear-mongering is already suggested by the deep cathartic pleasure that we take in imagining world-wide cataclysmic destruction, repeatedly and in endless variation, as demonstrated by our insatiable appetite for apocalyptic films and novels and the new genre of cli-fi.[8] I do not believe that we can explain this away as disaster capitalism's effort to »transform apocalypse into exciting entertainment for the multitudes«, as Frederick Buell suggests (Buell 2010: 31). Our appetite for world-ending fiction long predates the disaster capitalism that Naomi Klein (2007) describes, which must itself be understood through our deep cultural identification with the apocalyptic mode.

Finally, it is notable that both Hulme and Buell point to the introduction of new scientific language, especially that of systems theory, as influencing the specific forms of apocalyptic discourse in the environmental movement and climate change in particular. Prophetic revelation is a key feature of the Judeo-Christian apocalyptic tradition, and today's augers are the scientists who have introduced us to non-linearity, tipping points, feedback loops, and other chaotic disruptions of our received models for change. That such models inform our cosmic—that is, spatial—understanding of the possibility of catastrophe is clear, but they also have eschatological—that is, temporal—implications.[9] If we now live the world in an apocalyptic mode informed by non-linearity, has this not also complicated our experience of time?

2 Apocalyptic Temporalities

Turning now explicitly to apocalyptic temporality, we find that one of its defining features is said to be its linear directionality, either guided by divine providence or driven by natural forces, toward a catastrophic end-point, a ›Judgment Day‹, beyond which all individual human judgment is irrelevant (Foust and Murphy 2009: 154). Alongside the spectacular destruction of the current world, the narrative »prophecies (directly or indirectly) a new world order«, and Judgment Day marks the passage into this new age, which is therefore also a new time.[10] Nevertheless, finer-grained distinctions are possible within this broad characterization, as Christina Foust and William Murphy have shown through their examination of climate press coverage. Foust and Murphy take up Stephen O'Leary's distinction between ›tragic‹ and ›comic‹ variations of the apocalyptic narrative and show that these correspond to distinct temporalities. The tragic narrative is fatalistic, marching unstoppably toward pre-ordained or necessary destruction regardless of human actions. The tragic therefore »promotes a view of time and human action as closed«.[11] When this narrative is adopted, time is shortened or accelerated in its rush toward an unambiguous end-point, and this unalterable plunge is often characterised in terms of feedback loops, tipping points, and the destabilization of phenological cycles. Such tragic narratives are often supported by analogies with examples from fiction or from the deep geological past, such as a volcanic extinction event 250 million years ago (Foust and Murphy 2009: 159). In other words, they are modelled on temporalities that are either counterfactual or incommensurable with our mundane linear conceptions of time. The tragic narrative is reassuring in its own way precisely because there is nothing to be done, and we can resign ourselves to letting the course of things unfold—nothing to be done, that is, except to repent for our role in the corruption of the present (Foust and Murphy 2009: 161–162). The new world that succeeds final judgment will be purified of all corruption because it will continue on in its own pristine eternity, without us.

By contrast with this tragic narrative, although often mixed with it, the comic perspective is more optimistic about the prospects for human agency to influence the course of events. Things are headed badly due to our mistakes, and these can be righted. While we may not avoid disaster entirely, we might avoid the worst. The temporality of the comic, then, is open, and the point toward which it aims is less determinate. Rather than rushing toward a tipping point beyond which all human intervention is impossible, the comic mode slows things down long enough to give us time to think and to act. Within this narrative structure, then, we can hold on to the distinction between crisis and catastrophe (Foust and Murphy 2009: 160). Since Foust and Murphy approach the issue of apocalyptic rhetoric through the lens of its communicative and political efficacy, their proposal is for »communication scholars and climate scientists« to work together »on the difficult task of providing appropriate perspectives toward time«, perspectives that will encourage audiences to see climate change as urgent but manageable (Foust and Murphy 2009: 163). For them, this means promoting the »comic« perspective.

Nevertheless, this sorting into tragic and comic modes remains superficial until we contextualise it within the temporal structure of the full apocalyptic narrative. Whereas the tragic mode destabilises and accelerates time toward the definite moment when duration gives way to eternity, the comic mode aims to maintain a regular pace, a continuity with lived time, long enough to prepare us for judgment. But judgment still comes, in the sense that we must pass through a crisis for which we are as yet unprepared, and that maintains its full catastrophic potential depending on what we do next. On the comic mode, the world still passes into a new age and a new time, even if this is a new age in which we might still have a place. But this new age, this new time, and this new place are as yet unimaginable.

In insisting on the moment of crisis as a temporal hinge, as the turning point between ›our‹ time and a time to come, I am borrowing an insight from political theorist Ben Jones, who examines the appeal of Christian apocalyptic thinking for secular political theorists. Jones focuses on the strand of Christian thinking that he terms »cataclysmic apocalyptic thought«, exemplified by the Book of Revelation among other texts, that »identifies crisis as the path to the ideal society« (Jones 2017: 2). On this view, crisis is not to be avoided but rather welcomed since it is the only path that can wipe away the current state of corruption and replace it with lasting utopia (Jones 2017: 3). The truly apocalyptic crisis, then, is the final crisis, the one that installs us in a time beyond all crises. And this leavens our everyday struggles, here and now, with transcendent significance, insofar as they are moments of the larger progression toward final purification; we may be losing the local struggle, but we are still on the winning side of the cosmic battle (Jones 2017: 5; Skrimshire 2014: 239). My suggestion is that *our cultural fascination with fictional apocalyptic narratives is less a manifestation of our desire for our own destruction than our yearning for this transcendent significance; we are ready, in our heart of hearts, to wipe the world away and start again, even at the risk that we might be wiped away with it.* In the Christian version of this narrative, of course, the crisis and its aftermath unfold under the guidance of divine providence, and we need only have faith in this. Secular versions proceed without this safety net or try, like Marxism, to replace it by other mechanisms. In any case, the way that we live the apocalyptic narrative today is through our deep pleasure at the prospect of leaping into an unimaginable world and a new age without any guarantees of survival—and, importantly, without any unpaid debts to the past.

The radicality of this image of time follows from the unique moment of judgment, which is precisely a singular break where time folds, dehiscing into the old that is washed away and the ideal future to come. This returns us to Kant's late essay, »The End of All Things«, where he calls attention to the strange temporality of Judgment Day as the hinge between time and eternity, which both horrifies and attracts us with the full force of the sublime (Kant 1996). But for Kant this is a transition between the happening of events under the conditions of time, on the one hand, and an eternity in which nothing can come to pass, on the other, a situation that cannot be rationally comprehended but is to be understood according to the moral order of ends. Even if the new age postulated by climate apocalypticism does not altogether escape from time, as an incorruptible ideal it nevertheless retains a certain aura of the eternal or at least the indefinite openness of ›future generations‹. Judgment Day is always a selection, a differentiation of the corrupt from the pure, whether, along with

Kant, we place time on the side of corruption or instead allow for the possibility of a utopian time, whatever this might mean. Consequently, apocalyptic thinking relies on a linear sorting into two incommensurable but internally homogenous series of time, with the moment of judgment as their transition. As with Kant, it is the eternal or the utopian moment that remains sublime, unthinkable—and transcendent. But perhaps the sublime element of the unthinkable future is not a Judgment Day that opens onto eternity. In fact, does not the opening of the horizons of deep time, both past and future, transform our very experience of lived time? This new temporal sublime is neither a mathematical infinity of uncountable moments, nor a Judgment Day that closes time, but instead the anachronistic interruption of lived time by inconceivable durations that outstrip the imagination. This elemental sublime is covered over when the past is levelled into equivalent and interchangeable moments by which we extrapolate a manageable future.

3 The Singular Present and Spiralling Time

The suggestion I have been developing here is that our contemporary apocalypticism remains fundamentally eschatological, that it embraces crisis as a Judgment Day that marks the hinge between our corrupt present world and a new dawn, even or especially when this eschatological frame is not consciously or explicitly theological. It is this basic narrative that has underwritten environmentalism since at least *Silent Spring*, despite the modifications that it has undergone in the light of new technologies and shifting political contexts.[12] This narrative justifies itself in terms of our ethical obligations toward the future, and yet it assumes a figure of time that conceals our ethical obligations—not only toward the future, but also toward the past and present. To see why this is so, we must first recognise that the apocalyptic figure of time participates in what Jean-Luc Nancy (2015) has termed the »equivalence of catastrophes«. Nancy describes our global ecotechnical situation as an ever-expanding entanglement of interdependencies between innumerable systems—political, military, industrial, financial, logical, natural, and so on. Because of these interdependencies, catastrophes such as the 2011 disaster at the Fukushima Daiichi Nuclear Power Plant are uncontainable in their effects. But the deeper catastrophe, as Nancy argues, is the general equivalence that makes the interdependence of systems possible in the first place, namely, the levelling of all measures into a common denominator that facilitates translation across domains. This general equivalence inspires a proliferation of means and ends without orientation toward any final end or ultimate goal other than their own continued expansion and proliferation. It is this loss of any ultimate sense or direction that Nancy has called the »end of the world«.[13] Our constant awareness of the possibility of our own self-destruction stands in place of any final end as the secret fulfilment of the levelling of time into a homogenous continuum.[14] The operations of this catastrophe of equivalence can be traced in those approaches to sustainability that extrapolate from deep-past trends to predict and manage far-future scenarios, thereby tacitly assuming that our obligation toward the future is to ›sustain‹ the world in a state that resembles as closely as possible our present.[15]

Nancy points out that the absence of any end or goal for our ecotechnical interdependencies apart from their own self-perpetuation traps us in a cycle of planning and management of the future in general, and the extrapolation of the past to calculate the future demonstrates the sway of this general equivalence in our understanding of time, since each chronological present moment is substitutable and exchangeable for every other. Apocalyptic thinking presupposes and reaffirms this equivalence, since the only decisive interruption of the linear and calculable equivalence of ›nows‹ is by a Judgment Day that ends time as we know it, and precisely a Judgment Day that is itself the consequence of the proliferation of catastrophic equivalence—that is, one we have brought upon ourselves.

To break with this levelling of time requires recognising the *non-equivalence* of the unique and non-substitutable events and moments that compose our lives, moments that cannot be exchanged precisely because of their entanglement in the plexities of the past and future. To recognise this singularity of every moment deepens our respect for the present, understood not as an immediate or ephemeral ›now‹, but rather as the time of manifestation in which someone or something, always singular and incommensurable, presents itself. In Nancy's words, »What would be decisive, then, would be to think in the present and to think the present. No longer the end of ends to come, or even a felicitous dispersion of ends, but the present as the element of the near-at-hand«.[16] The singularity of what appears in the non-substitutable present demands from us an attention and respect, an esteem for the inestimable (Nancy 2015: 39–40). Wendell Berry, the leading proponent of agrarian ideals in the United States, expresses what may be a parallel sentiment when he writes that »We are always ready to set aside our present life, even our present happiness, to peruse the menu of future exterminations. If the future is threatened by the present, which it undoubtedly is, then the present is more threatened, and often is annihilated, by the future« (Berry 2015: 174).

Nancy's perspective allows us to glimpse the ethical implications of our figure of time. Yet, for a genuine encounter with the deep temporal sublime, it is necessary to extend Nancy's insights concerning the inestimable present. I suggest here two directions that this extension might take. First, the explosion of our temporal horizons far beyond the limits of human history, considered by Hume or Kant, and the parallel opening of a deep temporal future that continues beyond human extinction, confront us with the fact that our personal and historical temporalities are entangled and shot through with anachronistic and incompossible durations—those of our evolutionary history, for example, and, further still, of our own elemental materiality. The experience of the deep temporal sublime is characterised precisely by its incommensurability with the narrative structures of personal and cultural history, by the vertigo of losing all common markers and measures. This testifies to our entanglement in a past that was never our own possibility, never our own memory—an impossible and immemorial past. We live the abyssal unfathomability of deep time affectively and viscerally, in our heart of hearts, like a wedge driven through our lived experience of daily rhythms, our personal memories and anticipations, and the historical fabric of cultural events. Indeed, the very ›depth‹ of geological time is the bottomless free-fall into which it throws all markers and touchstones by which we orient ourselves within the temporal horizons of our world.[17] The schema of general equivalence is our unsuccessful attempt to repress this abyssal vertigo.

If we give up the effort to regiment time within general equivalence, then it becomes possible to consider our involvement, both material and symbolic, in incommensurable vectors of deep time: cosmic, geological, elemental, organic, evolutionary. From this perspective, our own corporeal materiality is liable to the immemorial past of the elements, while our organic lives enact the anonymous memories formed through diacritical evolutionary exchange with other species.[18] The encounter with the vertigo of deep time is thus the echo within us of evolutionary memory and the asubjective time of matter, which anachronistically interrupt our lived experiences of time from within. Insofar as these registers of time exceed the framing of human time and human worlds, the durations that precede our world cannot be disentangled from those that succeed it. A full accounting of the temporal sublime would therefore recognise the confluence of the immemorial past and future in its cosmic, geological, evolutionary, and organic trajectories, each with its own rhythms and durations.

Nancy sometimes speaks about the moment of presence as an interruption or suspension of continuity, a deferral of time's self-presentation, in favour of a relationship that demands a gesture or a response (Nancy 2017: 119, 121). Yet we see that what presents itself to us, here and now demanding our esteem and our response, may itself be of the past, or of the future. So we must recognise, as a second aspect of the deep temporal sublime, that a recovery of the present outside the calculable general equivalence of time also places us in an entirely different relationship to a past that is constitutive of our present possibilities and to futures that we do not plan or project.[19] Responsibility to the present therefore already involves us in the demands of justice for the past and the future. A profound example of how such temporal justice might be enacted is presented in the work of Potawatomi scholar Kyle Powys Whyte. Following on his critique of apocalyptic narratives of the Anthropocene, Whyte (2018) describes an experience of ›spiralling time‹ that expresses responsibility to the past and future within the present. Whyte notes that settler apocalyptic narratives, proposed as the effort of stopping »a dreaded future movement from stability to crisis« (Whyte 2018: 227), erase the legacies of colonial violence that have been experienced by many Indigenous people as repeated and ongoing apocalypses.[20] Furthermore, by seeking to liquidate the past and the present in a new beginning, the settler apocalyptic narrative imagines for itself an innocent future, one in which all obligations and debts for past and present colonial violence are assumed to be discharged. In contrast with such narratives of »finality and last-ness« (Whyte 2018: 236), Whyte describes Indigenous experiences of ›spiralling time‹ that maintain a continuous dialogue with one's ancestors and descendants. Whyte's account situates these experiences of time within a specific cultural context, yet he also invites non-Indigenous allies to engage in »counterfactual dialogue« and critical reflection on how the world that we inhabit today—that is, the world of colonial violence as well as climate change—is the dream and the gift of our settler ancestors, designed and constructed to »fulfill their fantasies of the future« and to »provide privileges to their descendants« (Whyte 2018: 237). Acknowledging that we are living the fantasy of our ancestors simultaneously opens a dialogue with our descendants, who pose to us the question of what kind of ancestors we ourselves will be, and what kind of world we will leave to those who follow. Counterfactual or fictional dialogue operates here not as an escape from our responsibilities to past and future, as we have seen in apocalyptic narratives, but rather as active affirmation of a spiralling of time that binds the manifestive present to the past that conditions it and the futures that it makes possible. In contrast with the calculative management of the future on the basis of the substitutability of homogenous times, and the linear finality of a Judgment Day that liquidates both past and future, such time spiralling interrupts and thickens the event of the present, in its inestimable singularity, with an anti-apocalyptic figure of temporal justice.

My aim here has been to reflect critically on the temporal framing of climate collapse narratives, taken here as paradigmatic of the environmentalism's apocalyptic rhetoric. I have argued that these narratives are structured by a moment of crisis, a Judgment Day, that grants transcendent significance to our present struggles through the promise of redemption in a new age. Although such narratives claim to be motivated by a sense of justice for the future, instead they conveniently serve to obscure legacies of past violence by levelling all moments of time into interchangeable and substitutable units. Temporal justice instead demands recognition of the non-equivalence of the present and its entanglement in incommensurable vectors of deep time, both past and future. As an alternative to apocalyptic narratives, I conclude with Kyle Whyte's example of temporal justice in the form of dialogue with our ancestors and descendants. Such ›time spiraling‹ affirms the temporal sublime while demanding critical reflection on our inescapable implication in the legacies of violence and fantasies of the future that constitute our singular present.

—Ted Toadvine, Penn State University, USA

References

W. Berry, Wendell, *Our Only World: Ten Essays*, Berkeley, 2015.
Brady, Emily, *The Sublime in Modern Philosophy: Aesthetics, Ethics, Nature*, Cambridge, 2013.
Brummett, Barry, »Premillenial Apocalyptic as a Rhetorical Genre«, *Central States Speech Journal*, Vol. 35, 1984, pp. 84–93.
Buell, Frederick, »A Short History of Environmental Apocalypse«, in: *Future Ethics: Climate Change and Apocalyptic Imagination*, ed. Stefan Skrimshire, London, 2010, pp. 13–36.
Clingerman, Forrest, and Kevin J. O'Brien, »Is Climate Change a New Kind of Problem? The Role of Theology and Imagination in Climate Ethics«, *WIREs Climate Change*, Vol. 8, 2017, pp. 1–10.
Derrida, Jacques, »No Apocalypse, Not Now: Full Speed Ahead, Seven Missiles, Seven Missives«, in: Jacques Derrida, *Psyche: Inventions of the Other*, Vol. 1, ed. Peggy Kamuf and Elizabeth Rottenberg, Stanford, 2007, pp. 387–409.
Fagan, Madeleine, »Who's Afraid of the Ecological Apocalypse? Climate Change and the Production of the Ethical Subject«, *The British Journal of Politics and International Relations*, Vol. 19, No. 2, 2017, pp. 225–244.
Fiskio, Janet, »Apocalypse and Ecotopia: Narratives in Global Climate Change Discourse«, *Race, Gender & Class*, Vol. 19, No. 1, 2012, pp. 12–36.
Foust, Christina and William O'Shannon Murphy, »Revealing and Reframing Apocalyptic Tragedy in Global Warming Discourse«, *Environmental Communication*, Vol. 3, No. 2, 2009, pp. 151–67.
Greenpeace International, »Horsemen of the Apocalypse Descend on Copenhagen«, *Scoop Independent News*, 16 December 2009, http://www.scoop.co.nz/stories/WO0912/S00466.htm (accessed 11/16/2018).
Gross, Lawrence, *Anishinaabe Ways of Knowing and Being*, New York, 2014.
Hulme, Mike, *Why We Disagree About Climate Change: Understanding Controversy, Inaction and Opportunity*, Cambridge, 2009.
Hulme, Mike, »Four Meanings of Climate Change«, in: *Future Ethics: Climate Change and Apocalyptic Imagination*, ed. Stefan Skrimshire, London, 2010, pp. 37–58.
Haller, Albrecht von, »Unvollkommenes Gedicht über die Ewigkeit / Uncompleted Poem on Eternity«, trans. Arnulf Zweig, *The Philosophical Forum*, Vol. 33, No. 3, 2002, pp. 304–311.
Hume, David, *A Treatise of Human Nature*, Vol. 1, ed. David F. Norton and Mary J. Norton, Oxford, 2007.
Hutton, James, *Theory of the Earth*, Edinburgh, 1788.
Jones, Ben, »The Challenges of Ideal Theory and the Appeal of Secular Apocalyptic Thought«, *European Journal of Political Theory* (OnlineFirst), 2017, pp. 1–24.
Kant, Immanuel, *The Critique of Judgment*, trans. Werner S. Pluhar, Indianapolis, 1987.

Kant, Immanuel, »The End of All Things«, in: Immanuel Kant, *Religion and Rational Theology,* trans. and ed. by Allen Wood and George di Giovanni, Cambridge, 1996, pp. 219–31.

Kant, Immanuel, »Observations on the Sublime and the Beautiful«, in: Immanuel Kant, *Anthropology, History, and Education,* ed. Robert B. Louden and Günter Zöller, Cambridge, 2007.

Killingsworth, M. Jimmie and Jacqueline Palmer, »Millenial Ecology: The Apocalyptic Narrative from *Silent Spring* to Global Warming«, in: *Green Culture: Environmental Rhetoric in Contemporary America,* ed. Carl Herndl and Stuart Brown, Madison, 1996, pp. 21–45.

Klein, Naomi, *The Shock Doctrine: The Rise of Disaster Capitalism,* New York, 2007.

Moo, Jonathan, »Climate Change and the Apocalyptic Imagination: Science, Faith, and Ecological Responsibility«, *Zygon,* Vol. 50, No. 4, 2015, pp. 937–48.

Nancy, Jean-Luc, *The Sense of the World,* trans. Jeffrey Librett, Minneapolis, 1997.

Nancy, Jean-Luc, *After Fukushima: The Equivalence of Catastrophes,* trans. Charlotte Mandell, New York, 2015.

Nancy, Jean-Luc, *The Possibility of a World,* trans. Travis Holloway and Flor Méchain, New York, 2017.

Nordhaus, Ted, and Michael Shellenberger, *Break Through: From the Death of Environmentalism to the Politics of Possibility,* Boston, 2007.

O'Leary, Stephen D, »A Dramatistic Theory of Apocalyptic Rhetoric«, *Quarterly Journal of Speech,* Vol. 79, 1993, pp. 385–426.

Playfair, John, »Biographical Account of the Late James Hutton, M.D.«, in: John Playfair, *The Works of John Playfair,* Vol. 4, Edinburgh, 1822, pp. 33–118.

ed. Skrimshire, Stefan, *Future Ethics: Climate Change and Apocalyptic Imagination,* London, 2010.

Skrimshire, Stefan, »The End of the Future: Hegel and the Political Ecology of Deep Time«, *International Social Science Journal,* Vol. 62, Nos. 205–206, 2011, pp. 325–338.

Skrimshire, Stefan, »Climate Change and Apocalyptic Faith«, *WIREs Climate Change,* Vol. 5, 2014, pp. 233–246.

Skrimshire, Stefan, »Deep Time and Secular Time: A Critique of the Environmental ›Long View‹«, *Theory, Culture, and Society* (OnlineFirst), 2018, 1–19.

Toadvine, Ted, »The Elemental Past«, *Research in Phenomenology,* Vol. 44, No 2, 2014, pp. 262–279.

Toadvine, Ted, »Biodiversity and the Diacritics of Life«, in: *Carnal Hermeneutics,* ed. Richard Kearney and Brian Treanor, New York, 2015, pp. 235–248.

Toadvine, Ted, »Our Monstrous Futures: Global Sustainability and Eco-Eschatology«, *Symposium,* Vol. 21, No. 1, 2017, pp. 219–230.

Toadvine, Ted, »Thinking After the World: Deconstruction and Last Things«, in: *Eco-Deconstruction: Derrida and Environmental Philosophy,* ed. Matthias Fritsch, Philippe Lynes, and David Wood, New York, 2018, pp. 50–80.

Whyte, Kyle Powys, »Indigenous Science (Fiction) for the Anthropocene: Ancestral Dystopias and Fantasies of Climate Change Crises«, *Environment and Planning E: Nature and Space,* Vol. 1, No. 1–2, 2018, pp. 224–242.

Wood, David, »On Being Haunted by the Future«, *Research in Phenomenology*, Vol. 36, 2006, pp. 274–298.
Wood, David, »Temporal Phronesis in the Anthropocene«, *Research in Phenomenology*, Vol. 47, 2017, pp. 220–227.
Wood, David, *Deep Time, Dark Times: On Being Geologically Human*, New York, 2018.

Notes

[1] Notable exceptions include Skrimshire (2011, 2018); Wood (2006, 2017, 2018); Whyte (2018).
[2] Hume (2007: 274–280); Kant (2007). Brady (2013) traces some of the early history of the temporal sublime.
[3] Such elemental examples of the dynamically sublime are also among the omens for Judgment Day (Kant 1996: 225).
[4] See in particular Playfair's description of his 1788 trip with Hutton to Siccar Point (1822: 80–81).
[5] See Toadvine (2017, 2018), where I develop in more detail the account summarily sketched here.
[6] Representative examples include Clingerman and O'Brien (2017), Fagan (2017), Fiskio (2012), Foust and Murphy (2009), Moo (2015), Skrimshire (2010, 2014).
[7] See Killingsworth and Palmer (1996); Fiskio (2012).
[8] See Toadvine (2017).
[9] See, for example, David Wood's brief remarks on ›tipping points‹ in Wood (2017: 223).
[10] Foust and Murphy (2009: 154; citing Brummett 1984).
[11] O'Leary (1993: 392; quoted in Foust and Murphy 2009: 154).
[12] See Killingsworth and Palmer (1996); Buell (2010).
[13] See Nancy (1997: 4–5).
[14] See Nancy (2015: 17–20).
[15] See Toadvine (2017, 2018).
[16] Nancy (2015: 37); see also Nancy (2017: 119–121, 122–126.
[17] See Toadvine (2014).
[18] I argue for these points, drawing on Merleau-Ponty, in Toadvine (2014, 2015).
[19] On the need to break from finality itself, i. e., ›from aiming, from planning, and projecting a future in general‹ and instead to work with ›other futures‹, see Nancy (2015: 37). On our ongoing responsibility to ›watch out‹ for the future, see Nancy (2015: 64n4). On the past that is constitutive of our present possibilities, see Wood on ›Constitutive Time‹ (2017: 7) and Toadvine (2014).
[20] As Anishinaabe scholar Lawrence Gross writes, »Native Americans have seen the ends of their respective worlds. […] Indians survived the apocalypse« (Gross 2014: 33; cited in Whyte 2018: 227).

Cross-cultural Explorations

Umweltethische Ansätze in der afrikanischen Philosophie*

Vor dem Hintergrund rasch fortschreitender Umweltbelastung und Naturzerstörung auf lokaler wie globaler Ebene, drängt sich die Umweltethik als eine vorrangige Aufgabe der gegenwärtigen Philosophie auf. Die wachsende Auseinandersetzung mit umweltethischen Themen hat in Afrika zur Bildung einer vielfältigen und lebhaften Philosophiedebatte geführt, bei der sowohl aktuelle Herausforderungen für Theorie und Praxis, als auch fundamentale Fragen zur Beziehung zwischen Mensch und Natur im Fokus stehen. Bei der Vielfalt der Herangehensweisen und Positionen zeichnet sich die Tendenz ab, traditionelle afrikanische Begriffe, Wertvorstellungen und Praktiken aufzugreifen und für die umweltethische Forschung fruchtbar zu machen.

Im Folgenden möchte ich die Arbeit von drei in verschiedenen afrikanischen Ländern tätigen Denkern vorstellen, die sehr unterschiedliche Perspektiven innerhalb der gegenwärtigen afrikanischen Umweltethik einnehmen, und sich durch Originalität und Aussagekraft auszeichnen. Die Positionen von Workineh Kelbessa, Segun Ogungbemi und Thaddeus Metz greifen auf kulturelle Traditionen afrikanischer Völker zurück, doch unterscheiden sie sich wesentlich sowohl in der Art und Weise, in der sie den Begriff Tradition verstehen und diese in der eigenen Forschung einsetzen, als auch hinsichtlich der methodischen Ausrichtung und Ziele ihrer Arbeit.

Anschließend werde ich einen kurzen Blick auf das Verhältnis zwischen Theorie und Praxis werfen. Anhand der katastrophalen Umweltlage im Niger Delta soll in diesem Abschnitt in exemplarischer Weise die Verflechtung regionalspezifischer Umweltprobleme in Afrika mit wirtschaftlichen, politischen und gesellschaftlichen Faktoren aufgezeigt werden. Angeschnitten wird die Frage, inwiefern die drei vorgestellten Herangehensweisen zur Umweltethik das Potential haben, bei ihrer Berücksichtigung afrikanischer Traditionen, die

Komplexität aktueller umweltethischer Herausforderungen in Afrika anzugehen.

Doch bevor ich auf die umweltethischen Reflexionen von Kelbessa, Ogungbemi und Metz eingehe, möchte ich noch einige Bemerkungen voranstellen, um die spezifischen Valenzen vorkolonialer Tradition im Kontext afrikanischer Philosophie abzuzeichnen. Denn traditionelles Gedankengut spielt nicht nur für die aktuelle afrikanische Umweltethik eine zentrale Rolle; vielmehr unterspannt die Auseinandersetzung mit dem Thema Tradition die gesamte Entwicklung der postkolonialen afrikanischen Philosophie.

1 Vorkoloniale Tradition zwischen Aufwertung und Idealisierung

Bereits in der Anfangsphase des afrikanischen Philosophiediskurses, die mit dem Ende der Kolonialzeit zusammenfiel, entwickelten sich, speziell im Rahmen der intellektuellen Strömungen der Négritude und der Ethnophilosophie, gewisse Haltungen und Tendenzen im Umgang mit der Tradition, die ich anreißen möchte, da sie bis heute die afrikanische Philosophie und mitunter auch die Umweltethik nachhaltig prägen.

Im Rahmen der antikolonialen literarischen Bewegung der Négritude hatte die Aufwertung der afrikanischen Kultur der vorkolonialen Zeit eine ausgesprochen politische Bedeutung gewonnen. Die Bewegung entwickelte sich während der Dreißigerjahre des zwanzigsten Jahrhunderts in Paris, der Hauptstadt der französischen Kolonialmacht. Hauptakteure waren junge Intellektuelle aus unterschiedlichen Ländern Afrikas und der Diaspora, die ihre politische Botschaft durch die literarische Produktion, insbesondere in dichterischer Form, verbreiteten. Zentrales Thema war das Bewusstsein, dass die ausbeuterischen Praktiken von Kolonialismus und Sklaverei die Annahme bestimmter theoretischer Spekulationen voraussetzen, nämlich die einer angeblichen Unterlegenheit der Afrikaner, ihrer intellektuellen Fähigkeiten, ihres Glaubens und ihrer Kultur. Dieses von den westlichen Kolonisatoren konstruierte, verzerrte Bild vom »primitiven wilden Afrikaner«, das als Rechtfertigungs- und Legitimierungsgrund für das systematische und tiefgreifende Unrecht des Kolonialismus gedient hat, galt es in der Négritude zu kontern.

Dabei wird die biologische Eigenschaft der dunklen Hautfarbe, die das Label »Négritude« stolz verkündet, zum Sinnbild für die kollektive Identität der Kolonisierten, deren Würde es zu rehabilitieren gilt. Es sei wohlbemerkt, dass die aus der gemeinsamen Erfahrung der Kolonisation erwachsene Vorstellung der kollektiven Identität der Menschen afrikanischer Abstammung auf dieselben dichotomischen Muster zurückgreift, welche die rassistischen Spekulationen der Kolonisatoren genährt hatten. Bei ihrer Gegenüberstellung von »Schwarz« und »Weiß«, Gefühl und Rationalität, verfolgten die Vertreter der Négritude unterschiedliche Ziele. Bei manchen ging es um Protest: In einer Dichtung, die von polemischen Tönen, expressionistischen Stilmitteln und emanzipatorischer Wut charakterisiert ist, wurde das Bild des »Wilden« sogar sarkastisch gepriesen und provokativ inszeniert – zum Beispiel in den ausdrucksstarken Versen von Aimé Césaire in *Cahier d'un retour au pays natal*. Jenseits der Provokation wurde jedoch die Idee einer grundlegenden, essentiellen Differenz zwischen »Weißen« und »Schwarzen« auch ernsthaft postuliert, vorzüglich in der Philosophie von Léopold Sédar Senghor. Das hat Spuren in der Entwicklung des afrikanischen Philosophiediskurses hinterlassen, und auch im Bereich der Umweltethik lassen sich noch Positionen finden, die eine verallgemeinernde und essentialistische Darstellung des »afrikanischen Denkens« liefern.

Zwischen Entfremdung und Sehnsucht nach einer originären, unverfälschten afrikanischen Identität rekurrierte in der Négritude aber auch ein anderes Thema, nämlich die Rückkehr in einen als mythischen Ort dargestelltes Afrika, in dem Menschen, Pflanzen- und Tiere in einem harmonischen Kreislauf vereint sind. Der nostalgische Rückblick auf dieses idealisierte Bild des vorkolonialen Afrikas als ein verlorenes Paradies ist eine Haltung, die in der aktuellen afrikanischen Philosophie und speziell auch in der Umweltethik stark vertreten ist.

Das ist auch der philosophischen Strömung der Ethnophilosophie zu verdanken, die einen einschlägigen Einfluss auf den postkolonialen philosophischen Diskurs in Afrika gehabt hat. Auch bei dieser Bewegung, die sich auf afrikanischem Boden entwickelte, spielt der Bezug auf Tradition eine entscheidende Rolle. Denn das Forschungsvorhaben der Ethnophilosophie besteht darin, das mündlich tradierte vorkoloniale Gedankengut zu sammeln, zu deuten und zu systematisieren. Die Veröffentlichung von *La Philosophie Bantoue* (1945) von Placide Tempel markierte – durch die erstmalige Würdigung afrikanischen Gedankenguts als Philosophie – den Beginn der ethnophilosophischen Forschung, die wenigstens bis in die Sechzigerjahre als eine vielversprechende, genuin afrikanische Art des Philosophierens verstanden und praktiziert wurde.

Die grundsätzlich rückblickende Herangehensweise der Ethnophilosophie wurde unter anderem von dem Philosophen und ehemaligen Kulturminister von Benin, Paulin Hountondji aus der Elfenbeinküste, kritisiert. Obgleich Hountondji die Bedeutung vorkolonialer Traditionen und ihrer philosophischen Ausarbeitung anerkennt, warnt er davor, Ethnophilosophie als ›die afrikanische Philosophie‹ zu begreifen: Durch die ausschließliche Ausübung kulturhistorischer Rekonstruktionen würden sich die afrikanischen Denker in einer zeitlosen geistigen Nische eingrenzen. Dabei würden sie sich selbst der Möglichkeit entbehren, die brennenden aktuellen Probleme Afrikas theoretisch anzugehen und an den internationalen Philosophiedebatten teilzunehmen (Hountondji 1994: 47).

Die nachhaltige Wirkung von Ethnophilosophie und Négritude lässt sich noch heute in der afrikanischen Philosophie spüren. Zum einen wird Ethnophilosophie noch praktiziert, durch nostalgische Rekonstruktionen einer Vergangenheit, deren dynamische Historizität in eine zeitlose und immerwährende Dimension gerückt wird. Zum anderen umfasst die gegenwärtige afrikanische Philosophie eine Vielfalt von thematischen und methodischen Ausrichtungen, die ein kritisches Herangehen an die Tradition und zugleich ein ausgeprägtes Bewusstsein für das philosophische Potential afrikanischer Kultur aufweist.

Das ist mitunter auch im Bereich der Umweltethik der Fall. Denn die Suche nach kulturell eigenen Alternativen zum anthropozentrischen und instrumentellen Naturverständnis – das auch Umweltethiker aus westlichen Ländern selbstkritisch als eine wichtige theoretische Voraussetzung für die Umweltkrise erkannt haben – erfolgt in der afrikanischen Umweltethik vor dem Hintergrund eines Diskurses, welcher den Bezug zur Tradition als identitätsstiftenden Faktor nach der traumatischen Zäsur der Kolonialerfahrung grundsätzlich aufwertet. Dabei soll es nicht überraschen, dass das in westlichen Ländern verbreitete Verständnis der Natur als unterworfen und grenzenlos verfügbar oft das Objekt einer harschen Kritik wird, bei der theoretische und geschichtliche Zusammenhänge mit der ausbeuterischen Haltung des Kolonialismus gezogen werden.

Ein bedenklicher Zusammenhang zwischen Kolonialismus und Umweltethik wird von Polycarp Ikuenobe hervorgehoben: Während die Afrikaner von den Kolonisatoren dazu indoktriniert wurden, ihre Glaubens- und Wertvorstellungen – als primitiv und irrational gebrandmarkt – aufzugeben und sich einen auf Gewinnmaximierung orientierten Umgang mit Tieren und Umwelt anzueignen, werden sie heute von westlichen Umweltschützern über die intrinsischen Werte der Natur belehrt. Ikuenobe bemerkt:

Contemporary environmentalists are *again* coming to Africa like the European colonialists did before them, with what appears to be a ›new‹ set of environmentalist ideas to save Africa and their environment. Yet these ›new ideas‹ represent the old traditional Africans' moral attitudes or ideas of conservation, which were not only rejected and destroyed by colonialism, but were also the bases for the negative stereotypes about Africa and Africans as undeveloped and uncivilized. It is now obvious that the models of development brought by European colonialists to Africa, which led to the destruction of the environment, are to a significant degree, wrong (Ikuenobe 204: 19).

Die Autoren, um die es im Folgenden geht, engagieren sich bei ihrer Arbeit für eine Aufwertung traditionell afrikanischen Gedankenguts, die den Anforderungen einer aktuellen Umweltethik und ihrem Anspruch auf Allgemeingültigkeit angemessen ist.

Die vertiefte Auseinandersetzung mit sozio-kulturellen und religiösen Praktiken afrikanischer Völker impliziert nicht notwendigerweise eine weltfremde nostalgische Haltung, sondern kann durchaus mit dem Bewusstsein für wirtschaftliche und ethische Konflikte vereint, den Herausforderungen einer zeitgemäßen Umweltethik gerecht werden. Exemplarisch für diese Ausrichtung ist die Arbeit von Workineh Kelbessa aus Äthiopien.

2 Workineh Kelbessa: die Oromo und der Wert indigener Naturethik

Während seiner langjährigen Forschungstätigkeit hat Workineh Kelbessa, Professor in Addis Abeba, dem kuschitisch-sprechenden Volk der Oromo – der größten ethnischen Gruppe Äthiopiens – umfangreiche Studien gewidmet. Der Regionalstaat Oromia liegt zentral in der Bundesrepublik Äthiopien, grenzt an Sudan und Somalia und erstreckt sich nördlich bis zum abessinischen Hochland, südlich bis nach Kenia. Es handelt sich um ein fruchtbares Gebiet, reich an Wasser, Rohstoffen und anderen natürlichen Ressourcen; was vermutlich auch einer der Gründe für die Unterdrückung und Ausbeutung, seitens auswärtiger sowie äthiopischer Akteure war, welche die Oromo-Geschichte geprägt haben.

Die Studien von Workineh Kelbessa liefern in erster Linie eine systematische und reflektierte Rekonstruktion des traditionellen Umgangs der Oromo mit der natürlichen Umwelt (Kelbessa 2001, 2010). Allerdings begrenzt sich Kelbessa nicht auf die Rekonstruktionsaufgabe, sondern untersucht auch die Spannungen, die sich ergeben, wenn man die traditionellen Werte und Praktiken der Oromo im konkreten Kontext globalisierter Dynamiken situiert. Dabei lässt sich eine gewisse Skepsis gegenüber romantisierenden Darstellungen traditioneller afrikanischer Kultur herauslesen. Seine Forschungsmethode kombiniert literarische Quellen aus Geschichte, Umweltethik und Recht mit Studien vor Ort, bei denen Kelbessa die Einsichten der Oromo zur Natur und Umweltethik sammelt und in eigene Beobachtungen integriert.

Aus Kelbessas Arbeit geht hervor, dass die Haltung der Oromo zur Natur stark normativ geregelt ist. Zuerst einmal gilt die Vorstellung, ganze Spezies aus der Pflanzen- und Tierwelt durch menschliches Einwirken auszurotten, als verpönt. Dies wird am Beispiel des Umgangs mit Wildtieren deutlich. Das Verhalten beim Jagen ist eingegrenzt durch eine Vielzahl an Regeln, die dazu dienen, die Reproduktionsfähigkeit der Wildtiere zu schützen: Sozial verachtet wird das Töten von läufigen und von schwangeren Tieren, sowie von Müttern mit ihren Kälbern.

Kelbessa führt weitere einleuchtende Beispiele der konsequenten Umsetzung dieses Prinzips auf. Wenn die Oromo eine Gruppe von Wildtieren jagen, sollen sie immer wenigstens zwei Exemplare am Leben erhalten: ein Männchen und ein Weibchen, um die Kontinuität der Spezies zu gewährleisten. Dieser Grundsatz gilt auch für die Pflanzenwelt, und begründet beispielsweise das Verbot, nie alle Äste eines Baumes abzuschneiden.

Die Vielzahl von Geboten und Verboten, welche den Umgang mit Flora, Fauna, Wasser und Boden regeln, wird von den Oromo unterschiedlich begründet. Auf einer ersten Ebene wird der Erhalt von Tier- und Pflanzenarten auf Grundlage von nutzenorientierten Überlegungen begründet: Das Wild ist eine Hauptnahrungsquelle für die Oromo; Bäumen spenden Holz, Früchte und Schatten.

Die Erhaltung der Biodiversität wird aber auch religiös begründet. So glauben die Oromo, dass Menschen, sowie Tier- und Pflanzenarten voneinander abhängig und in der gesamten Naturwelt miteinander verbunden sind, weil alles, was existiert, von einem Gott, *Waaqa*, erschaffen worden ist. Der harmonische Kreislauf der Natur wird durch ein normatives Prinzip der Gerechtigkeit, genannt *Safuu*, geregelt, das den Menschen verbietet, rücksichtlos mit anderen Schöpfungen und der Umwelt umzugehen. Bemerkenswert ist dabei, dass *Safuu* eine vom Nutzten für die Menschen unabhängige Existenzberechtigung der Tier- und Pflanzenwelt begründet: »The Oromo believe that it would be wrong to eliminate any species created by Waaqa« (Kelbessa 2010: 101).

Gemäß *Safuu* darf weder eine Spezies ausgerottet, noch einzelne Individuen aus der Tierwelt gequält oder unbedacht getötet werden. So fordert das von *Waaqa* erschaffene ethische Prinzip sowohl den Erhalt von Biodiversität als auch die Minimierung des Leidens von Lebewesen.

Bei seiner Forschung berücksichtigt Kelbessa auch den Umgang der Oromo mit Konflikten, die im Alltagsleben zwischen idealen Werten und konkreten Bedürfnissen entstehen. Wenn die Interessen von Tieren und Menschen kollidieren, zum Beispiel wenn Igel und Affen die Ernte beschädigen, oder Raubkatzen die Haustiere bedrohen, so ergreifen die Oromo lieber vorbeugende Maßnahmen und vermeiden die Tötung der unerwünschten Besucher (Kelbessa 2010: 99). Darüber hinaus ist auch der Umgang mit Boden und Wasser normativ geregelt. Ein komplexes System der Wassergewinnung- und Verteilung auf Mensch und Tier garantiert einen nachhaltigen Umgang mit der kostbaren Naturressource.

Darüber hinaus werden bestimmten Arten von Pflanzen und Tieren intrinsische Eigenschaften anerkannt. Dies gilt beispielsweise für Pferde, die als extrem intelligent gelten. Aufgrund ihrer Eigenschaften werden *Gollaa* (Sammelbegriff für Pferde, Esel und Kamele) von den Oromo besonders geschützt: Misshandlungen werden durch Sanktionen bestraft (Kelbessa 2010: 101).

Bei der Normierung des Umgangs mit Pflanzen wird, neben dem Glauben und der Nützlichkeit, auch der ästhetische Wert von Bäumen berücksichtigt (Kelbessa 2001: 41). Im Wertesystem der Oromo findet sich noch eine weitere Begründung für den nachhaltigen Umgang mit der Naturwelt, nämlich der Glauben, dass nicht nur gegenwärtige Bewohner der Erde, sondern auch zukünftige Generationen ethisch berücksichtigt werden sollen: »The Oromo believe that the present generation has responsibilty to pass on natural resources in good order to a future generation. That is why the Oromo are concerned with the health and peace of the environment and its inhabitants« (Kelbessa 2010: 77).

Zusammenfassend fordern die umweltethischen Grundsätze der Oromo den Erhalt der Biodiversität, die Minimierung von Schädigungen an anderen Lebewesen, sowie einen nachhaltigen Umgang mit den Naturressourcen. Objekt umweltethischer Interessen sind Menschen, Tiere, Pflanzen und Umwelt sowie vergangene, gegenwärtige und zukünftige Generationen.

Ein bedeutsamer Aspekt der Reflexion von Kelbessa ist, dass er kein einseitig positives Bild der Tradition liefert. Die kritische Haltung gegenüber idealisierten Darstellungen indigener Kultur zeigt sich in seinem Bemühen, die Schattenseiten afrikanischer Tradition aufzudecken. Dabei weist Kelbessa auf diskriminierende soziale Rollenmuster, beispielsweise auf die unterlegene Stellung der Frau innerhalb der Oromo Gesellschaft, hin (Kelbessa 2010: 163).

Darüber hinaus geht Kelbessa auf die Frage nach möglichen Konflikten zwischen unterschiedlichen ethischen Werten, insbesondere zwischen umweltethischen Zielen und dem fundamentalen Recht auf Nahrung und dem Wohlergehen der Menschen, ein. Als ein möglicher Weg, Armut und Hungersnot zu bekämpfen, zeichnet sich die Einführung moderner Anbautechnik ab. Diesbezüglich liefert Kelbessa eine nüchterne Analyse: Moderne Produktionstechniken werden, wie die Erfahrung zeigt, allzu oft unbedacht eingesetzt; dabei werden weder Biodiversität und Nachhaltigkeit, noch die indigenen Werte, Wissen und Bedürfnisse berücksichtigt. Denn die Anwendung von Monokulturen, mechanisierter Agrartechnik und chemischen Vertilgungsmitteln hat in Äthiopien zu Bodenverarmungen, Erosionen und massiver Entwaldung geführt. Darüber hinaus stehen Bauern und Hirten den neuen Anbaumethoden, die sich ihrer tradierten Agrarpraxis und ihrem Wertesystem entgegenstellen, skeptisch gegenüber. Weiterhin wird die indigene Bevölkerung zunehmend ihres Lebensraums beraubt, der gegen ihre Vorstellungen, Interessen und ihr Selbstbestimmungsrecht tiefgreifend umgestaltet wird.[1]

Kelbessa plädiert für eine Integration von traditionellem Wissen und moderner Technik, welche die Expertise der indigenen Bevölkerung, ihr Wissen über die lokale Umweltbeschaffenheit und den nachhaltigen Umgang mit der Natur sowie ihre Interessen berücksichtigt. Neue Wege der Produktionsimplementierung seien nur dann eine sinnvolle Entwicklung, wenn sie nicht auf Kosten von Nachhaltigkeit und Gerechtigkeit erfolgen.

Die Ergebnisse, die Kelbessa erlangt, sind in vielerlei Hinsicht von Bedeutung. Neben einer detaillierten Darstellung des umweltethischen Wissens der Oromo liefert er aufschlussreiche Einsichten in das Potential einer traditionellen Kultur im Lichte gegenwärtiger umweltethischer Probleme. Zwar lässt sich aus der Umweltethik der Oromo kein einheitliches Prinzip universeller Gültigkeit bestimmen. Ob man eines solchen Prinzips bedarf, bleibt allerdings eine offene Frage. Zweifellos bildet die kulturell verwurzelte Achtung der Oromo für die Naturwelt eine wertvolle Voraussetzung für die Entwicklung umweltfreundlicher Strategien und ihrer effektiven Anwendung auf lokaler und regionaler Ebene und ist möglicherweise auch ein Beispiel für andere Kulturen, das Anerkennung verdient.

3 Das umweltethische Prinzip von Segun Ogungbemi

Eine sehr unterschiedliche Herangehensweise an Tradition und Umweltethik vertritt Segun Ogungbemi aus Nigeria.

Ausgangspunkt seiner umweltethischen Reflexion in *An African Perspective on the Environmental Crisis* ist eine Untersuchung der Ursachen der gegenwärtigen ökologischen Krise. Dabei fokussiert er, ohne die globale Dimension auszuklammern, auf die spezifische Umweltlage im subsaharischen Afrika. Die unterschiedlichen Faktoren, die zur Umweltbeschädigung in Afrika beitragen, unterteilt Ogungbemi in drei Kategorien: erstens, Armut und Unkenntnis; zweitens, moderne Produktionssysteme und drittens, politische Konflikte, die oft mit internationalen wirtschaftlichen Interessen einhergehen.

Charakteristisch ist für Ogungbemi der nüchterne Blick, mit dem er traditionelle afrikanische Lebensweisen betrachtet. Obgleich Ogungbemi die umweltfreundliche Orientierung tradierter Naturauffassungen anerkennt, stellt er fest, dass auch traditionelle Gesellschaften ihren Beitrag zur Umweltkrise leisten, indem sie Bräuche praktizieren, die umweltschädlich sind. Hierzu zählen beispielsweise das Verbrennen von Gebüsch bei der Jagd, die ausschließliche Verwendung von Brennholz als Energiequelle sowie die Verschmutzung von Gewässern bei alltäglicher Hygiene. Diese umweltschädigenden Praktiken sind grundsätzlich dem Mangel an sanitären und energetischen Infrastrukturen, sowie dem mit Armut einhergehenden Unwissen verschuldet.

Andererseits sieht Ogungbemi im Streben vieler Afrikaner nach westlichen Konsumstandards sowie in der massiven Produktion von Exportgütern wichtige Ursachen für Umweltverschmutzung und Bodenverarmung.

Ein besonderes Augenmerk schenkt Ogungbemi dem Zusammenhang zwischen Umweltschäden und Politik. Hervorgehoben werden zum einen die schädigende Wirkung von militärischen Interventionen und Aufrüstungsprogrammen, zum anderen die Kurzsichtigkeit und Korruptionsbereitschaft von manchem politischen Leader in Afrika. Durchaus bedenklich ist in diesem Zusammenhang die zunehmende Praxis, toxische Abfälle, die aus westlichen Ländern stammen, an den afrikanischen Küsten und auf afrikanischem Boden zu entsorgen. Dieses Problem, das jüngst auch von Workineh Kelbessa angezeigt wurde,[2] bringt die enge Verflechtung zwischen Umweltbelastung, Menschenrechtsverletzung und Korruption zu Tage.[3]

Die Analyse der spezifischen Umweltprobleme Afrikas und deren Ursachen bildet den Rahmen, in dem Ogungbemi neben praktischen Lösungsvorschlägen, wie dem Ausbau von Solarenergie, auch die theoretische Frage nach einem normativen umweltethischen Prinzip angeht. Sein Ziel ist es nicht, wie Kelbessa, die traditionelle Umweltethik eines bestimmten afrikanischen Volkes im Detail zu untersuchen, sondern ein ethisches Prinzip zu finden, das im afrikanischen Gedankengut wurzelt und zugleich dem Anspruch auf Allgemeingültigkeit gerecht wird. Um dies zu gewährleisten, greift Ogungbemi zunächst auf ein gemeinsames ethisches Prinzip traditioneller afrikanischer Gesellschaften zurück: nicht mehr aus der Natur zu nehmen, als man braucht. Dieses Prinzip wird von Ogungbemi zum Kern einer in metaphysischen und religiösen Naturvorstellungen wurzelnden afrikanischen Care-Ethik erklärt. Genauere Merkmale und Begründungen einer solchen Care-Ethik werden von Ogungbemi nicht weiter ausgearbeitet; vielmehr wird dieses Prinzip als eine tradierte ethische Haltung postuliert: »In our traditional relationship with nature, men and women recognize the importance of water, land and air management. To our traditional communities the ethics of not taking more than you need from nature is a moral code« (Ogungbemi 2001: 266).

Nun stellt sich für Ogungbemi die Frage, ob die Umweltethik

traditioneller Gesellschaften, die er im oben genannten Prinzip resümiert, auf unsere zeitgenössische Situation anwendbar ist.

Im ethischen Gebot, nicht mehr aus der Natur zu nehmen, als man braucht, erkennt er einige Schwächen. Zuerst einmal zeichnen sich empirische Schwierigkeiten ab: Wer soll darüber urteilen, ob man mehr aus den Naturressourcen nimmt, als man braucht, und welche Sanktionen wären dafür denkbar? Schließich ist die Bestimmung dessen, was man braucht, ein äußerst schwieriges Unterfangen: »how do we know how much we need, given the nature of human greed?« (Ogungbemi 2001: 270) Hierbei möchte ich hinzufügen: Wie lässt sich das, was man benötigt, in einer kapitalistischen globalen Gesellschaft bestimmen?

Darüber hinaus würde ich grundsätzlich in Frage stellen, ob die Pflichten, die sich aus diesem Prinzip ableiten lassen, den aktuellen umweltethischen Herausforderungen, beispielsweise in Sachen Nachhaltigkeit, Biodiversität und Klimaschütz gerecht werden können. Denn das Verbot, aus der Natur nicht mehr zu nehmen als das, was man benötigt, steht nicht notwendigerweise im Einklang mit dem Gebot, so zu handeln, dass ein bestimmter Zustand der Natur erhalten bleibt.

Obgleich dieser Einwand von Ogungbemi nicht direkt adressiert wird, erkennt er, dass das traditionelle ethische Prinzip nicht ohne weiteres auf die zeitgenössische Umweltethik übertragbar ist. Daher sieht er die Notwendigkeit, die von ihm umrissene traditionelle afrikanische Umweltethik den aktuellen umweltethischen Anforderungen anzupassen.

Seinen Entwurf einer zeitgemäßen Umweltethik bezeichnet Ogungbemi als »ethics of nature-relatedness«. Dabei greift er auf die Idee der Relationalität zwischen Mensch und Umwelt zurück, die auch bei Kelbessas Darstellung der Umweltethik der Oromo, wie oben dargelegt, eine wesentliche Rolle spielt.

Allerdings unterscheidet sich Ogungbemis »nature-relatedness« vom traditionell afrikanischen Verständnis der Beziehung zwischen Mensch und Natur. Dort wird diese als ein harmonisches Verhältnis der gegenseitigen Abhängigkeit aller Entitäten beschrieben. Ogungbemi hingegen begreift diese Beziehung als ein asymmetrisches Verhältnis, bei dem die Abhängigkeit der Menschen von der Naturwelt in den Vordergrund tritt: Während die Natur wohl ohne Menschen weiter existieren könnte, sind die Menschen auf die Natur, und zwar auf einen bestimmten Zustand der Naturwelt, vollkommen angewiesen (Ogungbemi 2001: 270).

Die Tatsache, dass durch die Schädigung der Umwelt die Lebensqualität, ja sogar die Existenz der Menschen gefährdet werden kann, zeichnet sich als die fundamentale Begründung für eine nachhaltige Ethik ab. Der rücksichtvolle Umgang mit der Natur wird also letztendlich durch ein Prinzip begründet, das aus der Ethik der zwischenmenschlichen Beziehungen abgeleitet ist, nämlich durch das Verbot, Mitmenschen Schäden zuzufügen.

Es ist zu bemerken, dass Ogungbemi sich um eine auf breite Zustimmung abzielende Theorie bemüht: Er verzichtet auf religiöse Begründungen, um Allgemeingültigkeit zu gewährleisten.

Sein Ethikentwurf hat jedoch im Rahmen der afrikanischen umweltethischen Debatte seine Kritiker. Ojomo bemerkt: ›One major problem with Ogungbemi's ethics of nature-relatedness is that it bears little affinity or no semblance with African cultural experience and ontology‹ (Ojomo 211: 110). In der Tat bleibt unklar, was von der von Ogungbemi dargestellten traditionellen afrikanischen Care-Ethik, nach seiner Neuformulierung, erhalten bleibt. Diese entbehrt sowohl der religiösen und spirituellen Dimension als auch dem Bild eines harmonischen Zusammenhanges aller Entitäten miteinander und bewahrt möglicherweise nur das Gebot eines nachhaltigen Umgangs mit der Natur.

Fainos Mangena bemerkt, dass der von Ogungbemi geprägte Begriff nature-relatedness für die afrikanische Umweltethik von Bedeutung ist, bedenklich sei allerdings, dass in Ogungbemis Umweltethik den nicht-menschlichen Lebewesen lediglich ein instrumenteller Wert zugeschrieben wird (Mangena 2003: 35–36).

Ogungbemis Appell an die Vernunft, sowie seine klare Diagnose der Ursachen der Umweltkrise im subsaharischen Afrika bezeugen seine rationalisierende und weltnahe Herangehensweise. Umso überraschender wirkt folgende Anmerkung zum Thema Bevölkerungswachstum: »Generally, nature has a way of reducing population in the world. Recently, there was an earthquake in Egypt and many lives were lost. I believe when our population has reached an alarming situation, nature will invariably apply its brake and have a drastic reduction in our population growth rate« (Ogungbemi 2001: 269). Die Rede von einem Zusammenhang zwischen menschlichem Handeln und Naturkatastrophen mag zwar ein Leitthema der Umweltethik sein. Allerdings ist Ogungbemis Beschreibung dieses düsteren

Szenarios irritierend. Zum einen ist das Bild einer Natur, die aus einer bewussten Entscheidung heraus die Umweltbelastung durch Katastrophen regelt, bestenfalls als metaphorisch zu deuten; zum anderen ist die Annahme eines Zusammenhangs zwischen dem Wachstum der Bevölkerungsrate und dem Erdbeben in Ägypten etwas zu gewagt.

Einigen Schwächen zum Trotz hat die Arbeit von Ogungbemi das Verdienst, die Ursachen aktueller Umweltprobleme im Subsaharischen Afrikas eingehend zu analysieren und die Verflechtung der Umweltkrise in Afrika mit politischen und wirtschaftlichen Zuständen hervorzuheben. Zudem strebt er, angesichts der globalen Dimension der Umweltkrise, eine weltliche und allgemein zustimmungsfähige Umweltethik an.

Das Projekt einer Ethik, die auf traditionelle afrikanische Weltanschauungen fußt und zugleich auf religiöse Begründungen verzichtend, dem Anspruch auf Allgemeingültigkeit gerecht wird, verfolgt auch Thaddeus Metzt. Er tut das allerdings aus einer ganz anderen Perspektive heraus.

4 Thaddeus Metz: eine relationale Theorie des moralischen Status

Der Beitrag vom in Südafrika tätigen Philosophieprofessor Thaddeus Metz aus den Vereinigten Staaten fokussiert auf ein zentrales Thema umweltethischer Forschung, nämlich auf die Frage nach dem moralischen Status.

Einer Entität moralischen Status zuzuschreiben bedeutet, sie als moralisch berücksichtigungswürdig anzuerkennen, und zwar unabhängig von extrinsischen Fakten, beispielsweise von ihrer Nützlichkeit für die Menschen. Es gibt sehr unterschiedliche Auffassungen darüber, welche Eigenschaften für die Zuschreibung eines moralischen Status relevant sind. Anders als holistische Ansätze, welche die Natur als Ganzes als Träger von moralischem Status anerkennen, schreiben individualistische Theorien lediglich Individuen moralischen Status zu, typischerweise aufgrund intrinsischer Eigenschaften wie Rationalität oder Leidensfähigkeit. Thaddeus Metz nimmt sich vor, einen dritten Weg einzuschlagen und eine Alternative zu den dominanten Perspektiven des Holismus und Individualismus zu konzipieren, welche unseren moralischen Intuitionen besser gerecht wird.

In seiner relationalen Theorie bestimmt er als Kriterium für die Zuschreibung eines moralischen Status die Fähigkeit, Teil einer moralisch relevanten Beziehung mit Menschen zu sein. Abhängig davon, ob eine Entität nur das Objekt oder auch das Subjekt einer moralisch relevanten Beziehung sein kann, wird ihr ein voller oder nur ein partieller moralischer Status zugeschrieben. Innerhalb dieser Unterteilung werden wiederum unterschiedliche Grade der Relationsfähigkeit, und daher des moralischen Status, anerkannt. Es fragt sich zunächst, welche Art der Beziehung als moralisch relevant zu betrachten sei.

An dieser Stelle greift Metz auf traditionell afrikanisches Gedankengut zurück und macht das bekannteste ethische Prinzip des Subsaharischen Afrikas, Ubuntu, für seine Theorie fruchtbar. Nach einem weit vertretenen Verständnis, das auch der Arbeit von Metz zugrunde liegt, lässt sich Ubuntu wie folgt konturieren. Ubuntu ist eine moralische Eigenschaft, die entwickelt werden kann. Ein Mensch, der Ubuntu hat, identifiziert sich mit seinen Mitmenschen und verhält sich Anderen gegenüber solidarisch, kooperativ und hilfsbereit. In dem man das tut, fördert man das Wohl anderer, und damit auch das gemeinschaftliche Zusammenleben; zudem entwickelt man die eigene moralische Persönlichkeit; man wird also zu einem besseren Menschen. Damit ist Ubuntu zugleich deskriptive Eigenschaft und ethisches Prinzip, das in erster Linie die zwischenmenschlichen Beziehungen regelt.

Metz knüpft an das Konzept einer auf Identifikation und Solidarität beruhenden Beziehung an und macht daraus ein Kriterium des moralischen Status. Einer Entität wird moralischer Status zugeschrieben, wenn sie fähig ist, Teil einer Beziehung mit Menschen zu sein, die folgende zwei Kriterien erfüllt: Sie muss das Wohl des Anderen fördern und auf Solidarität und Identifikation beruhen. Anhand dieser zwei Kriterien lässt sich bestimmen, welche Entitäten moralischen Status haben, und in welchem Grad. Vollen moralischen Status genießen Entitäten, die das Subjekt einer solchen Beziehung sein können, die also fähig sind, sich mit anderen zu solidarisieren und ihr Wohlergehen zu fördern. Diese Fähigkeit schreibt Metz Menschen und, in geringerem Maß, auch bestimmten Affenarten wie Schimpansen und Gorillas zu (Metz 2011: 400).

Einen partiellen moralischen Status haben Entitäten, die nur das Objekt einer moralisch relevanten Beziehung sein können. Voraussetzung dafür ist, dass sie ein Leben haben, dessen Qualität gefördert werden kann. Das bedeutet, dass nicht lebendige Entitäten, mitunter Spezies und Naturressourcen, kein moralischer Status zugeschrieben werden kann.[4] Auch Pflanzen wird moralischer Status abgesprochen, weil sie kein im relevanten Sinne gutes Leben haben können.

Zudem muss eine zweite Bedingung erfüllt werden, nämlich die Fähigkeit, das Objekt von Solidarität und Identifikation zu sein. Das können Tiere in unterschiedlichem Maße. Laut Metz gibt es »animals that are capable of a better or worse life and so could be objects of a friendly relationship, but would be less able to be objects than other animals due to either their comparative lack of sentience or the dispositions of human nature, say, fish« (Metz 2011: 400).

Metz glaubt, mit seiner Ubuntu inspirierten relationalen Theorie eine Lösung zu bieten, die unseren moralischen Intuitionen entspricht und die kontraintuitiven Ergebnisse anderer Theorien des moralischen Status vermeidet. Darüber hinaus sei seine Konzeption von anderen Formen der Care-Ethik zu unterschieden, da sie durch das Hervorheben einer relationalen Fähigkeit den Vorteil bietet, weder Reziprozität noch das tatsächliche Bestehen einer Beziehung vorauszusetzen.

Bei genauerer Betrachtung ist Metzs Theorie allerdings nicht unproblematisch. Fraglich scheint mir die Art, in der Metz die Fähigkeit, das Objekt einer moralisch relevanten Beziehung zu sein, konturiert. Vorausgesetzt wird zum einen eine objektive Eigenschaft, nämlich »being able to live a good life« (Metz 201: 400); zum anderen die Fähigkeit des Objektes, Gefühle von Solidarität, Identifikation und Sympathie im Subjekt hervorzurufen. Inwiefern diese Gefühle von nicht-relationalen, objektiv vorhandenen Eigenschaften des Objektes abhängen, bleibt eine offene Frage. Denn die Unterschiede im moralischen Status, die Metz als hierarchisch auffasst, scheinen letztlich eine Funktion der Art und Weise zu sein, in der die Beschaffenheit einer Entität die menschlichen Neigungen anspricht. Das, was Metz als eine Fähigkeit des Objektes definiert, erweist sich schließlich als eine Präferenz des Subjektes der Beziehung. Folgende Formulierung bringt dies auf den Punkt: »a being can be the object of a friendly relationship insofar as characteristic human beings could think of it as part of a ›we‹, share its goals, sympathize with it« (Metz 2011: 394).

Das ist in vielerlei Hinsicht bedenklich. Abgesehen davon, dass »characteristic human beings« kulturell- und erfahrungsbedingt über ein und dieselbe Entität völlig unterschiedlich denken können, ist der normative Gehalt dieser Konzeption zweifelhaft.

Metz hat möglicherweise Recht, wenn er behauptet, dass sein Ansatz besser als andere Theorien in der Lage ist, Intuitionen bezüglich des moralischen Status zu erklären. Doch was Theorien des moralischen Status zu leisten versuchen, ist nicht die Frage, welche Entitäten wir intuitiv berücksichtigen, sondern welche wir berücksichtigen *sollen*. Die Umweltethik hat seit ihren Anfängen versucht, Intuitionen bezüglich nichtmenschlicher Entitäten zu hinterfragen, im Bewusstsein, dass diese manchmal, auf tradierte Konventionen beruhend, Leiden und Ungerechtigkeit fördern können.

Metzs Konzeption postuliert im Grunde die intuitive Empathie von Menschen als das Kriterium für eine hierarchische Skala des moralischen Status, welche dem Vorwurf von Willkür ausgesetzt ist.

Gegenwärtig gilt Metz als einer der produktivsten Forscher von Ubuntu. Sein Versuch, das Potential von Ubuntu in jede Richtung zu erkunden, und im Hinblick auf aktuelle ethische Fragestellungen anzuwenden, halte ich für durchaus legitim. In der Tat ist Relationalität ein zentraler Begriff afrikanischer Tradition, der für die Umweltethik fruchtbar gemacht werden kann. Es geht dabei jedoch um eine allumfassende Verbundenheit aller Entitäten, denen – unabhängig von ihren Beziehungen mit den Menschen – ein Eigenwert und ein Platz in der Welt anerkannt wird.[5] Diese Art von Relationalität – dessen Potential Metz nicht aufgreift – kann durchaus durch Ubuntu untermauert werden, beispielsweise durch die Erweiterung des normativen Prinzips von der Ebene der zwischenmenschlichen Beziehungen auf die gesamte Naturwelt. Einen solchen Ansatz wird zum Beispiel von Murove fruchtbar gemacht: Er legt die gesellschaftliche Dimension von Ubuntu als einen Mikrokosmos dar, welcher in kleinerem Maßstab die Verbundenheit der Weltordnung wiederspiegelt (Murove 2009).[6]

Während auf theoretischer Ebene eine rege Diskussion über afrikanische Umweltethik stattfindet, spielen sich auf dem afrikanischen Kontinent katastrophale ökologische Szenarien ab. Inwiefern die afrikanische Umweltethik diese konkreten Herausforderungen angehen kann, wird im Folgenden am Beispiel der Lage im Nigerdelta angerissen.

5 Der Fall Nigerdelta

Die im Süden Nigerias gelegene Mündung des Nigers umfasst eine Fläche von 70.000 Quadratkilometern und erstreckt sich über neun nigerianische Föderalstaaten. Es handelt sich um ein von Wasserläufen und Mangrovenwäldern durchzogenes Feuchtgebiet, reich an Fisch und Wasservögeln. Traditionell lebten die Bewohner des Nigerdelta von Landwirtschaft und Fischerei. Allerdings hat die Entdeckung von Erdöl in Oloibiri 1956 den Beginn einer drastischen Veränderung der Umweltlage im Nigerdelta markiert. Die intensive Ölexploration- und Gewinnung hat, durch das ständige Auslaufen von Erdöl in Wasser und Boden und das Abfackeln von giftigen Erdgasen, den natürlichen Zustand der Region enorm beschädigt: mit erheblichen Folgen für Flora, Fauna und die Lebensqualität der indigenen Bevölkerung. Die Bewohner des Nigerdelta sind ihres traditionellen Lebensunterhalts beraubt worden und leben in einer extrem verseuchten Umwelt.

Dabei erfährt der größte Teil der Bevölkerung nur die Nachteile des Ölgeschäfts, da sie kein Eigentumsrecht auf ihr Land geltend machen kann: durch den 1978 erlassenen und noch heute wirksamen *Land Use Act* wurde den Bewohnern der Besitz des Landes abgesprochen und dem Staat übertragen. Seitdem müssen die multinationalen Ölkonzerne die Nutzung des Landes nicht mehr mit Bürgern verhandeln (Ebeku 2002). Gegenüber diesen Angriffen auf ihr Land, ihr Leben und ihre Rechte ist die Bevölkerung nicht untätig geblieben. Durch zahlreiche Initiativen des friedlichen Protests haben die im Nigerdelta lebenden Ethnien versucht, ihre Rechte zu erkämpfen; doch seit der Tötung des Anführers der MOSOP Bewegung, Ken Saro Wiwa, und anderer acht Aktivisten des Ogoni Volkes im Jahr 1995 durch das Militärregime von Abacha, ist der Widerstand in Gewalt eskaliert. Während bewaffnete Guerillagruppen das Gebiet durchstreifen, gehören das Sabotieren von Pipelines und die Geiselnahme westlicher Mitarbeiter multinationaler Konzerne zur Tagesordnung.[7]

Der Fall Nigerdelta zeigt auf exemplarische Weise den Zusammenhang der Umweltkrise innerhalb und außerhalb Afrikas mit wirtschaftlichen, politischen und gesellschaftlichen Dynamiken. Es fragt sich, ob die umweltethischen Ansätze von Metz, Ogungbemi und Kelbessa Ressourcen bieten, um die Herausforderungen der Lage im Nigerdelta anzugehen.

Bezüglich Metzs Theorie bilden die von Ubuntu verkörperten

Werte von Kooperation, Solidarität und geteilter Menschlichkeit ein Gegenbild zum aktuellen Kampfszenario im Nigerdelta, geprägt von Ausbeutung, Unrecht und Korruption. Allerdings stellt uns die Lage im Nigerdelta vor die krude Tatsache, dass Solidarität und Identifikation leicht preisgegeben werden können, sobald wirtschaftliche Interessen im Spiel sind. Das Ubuntu-Ideal erfordert konkrete Spielräume für seine Verwirklichung, und diese sind, in einer Situation des tiefsten Unrechts, schwer gegeben. Sie zu gewährleisten wäre eine primäre politische Aufgabe.

Darüber hinaus bildet der Fall Nigerdelta ein konkretes, einleuchtendes Beispiel dafür, dass sämtliche Naturerscheinungen und Lebewesen, wenn ihre Umwelt beschädigt wird, dem Zerfall der gemeinsamen Existenz bestimmt sind. Gegenüber der Evidenz dieser allumfassenden Gemeinschaft greift Metzs Fokus auf die am menschlichen Standard gemessene Interaktionsfähigkeit möglicherweise zu kurz.

Bei Ogungbemis Theorie tritt die Verletzlichkeit des Menschen gegenüber dem ökologischen Zusammenbruch in den Vordergrund. Es ist kein Zufall, dass seine Analyse der Ursachen und Dynamiken der afrikanischen Umweltkrise auf die Lage im Nigerdelta besonders zutrifft, da Ogungbemi als Nigerianer das örtliche Geschehen erster Hand kennt (Ogungbemi 2010). Bezüglich des von ihm behaupteten traditionellen afrikanischen Prinzips, nicht mehr aus der Natur zu nehmen, als das, was man braucht, möchte man fragen, ob Erdöl – angesichts der Fülle an ethischen und umweltethischen Problemen, die mit seiner Gewinnung und Verwendung einhergehen, sowie der möglichen Alternativen – etwas ist, was die Menschheit wirklich benötigt.

Unter den drei vorgestellten Positionen setzt sich diejenige von Kelbessa am tiefsten mit tradiertem afrikanischem Gedankengut auseinander. Dabei entwickelt er einen originellen umweltethischen Ansatz, der neben dem Wert der nachhaltigen Umweltethik der Oromo als Kulturerbe auch den dynamischen Charakter der Tradition hervorhebt. So gewinnt für Kelbessa, wie für Ogungbemi, die Frage nach dem angemessenen Verhältnis zwischen Tradition und Erneuerung eine zentrale Bedeutung. Die Ausführungen Kelbessas zu den bestehenden Machtverhältnissen, zu einer bedachten Integration von traditionellem und technologischem Wissen, sowie zur Involvierung indigener Völker in die ihre Umwelt betreffenden Entscheidungsprozesse, sind auch im Hinblick auf das Nigerdelta aufschlussreich. Denn dort zeigt sich, dass bei der Lösung afrikanischer Umweltprobleme von der Berücksichtigung moralisch relevanter Fragen globaler Gerechtigkeit, den Rechten indigener Völker, der Verantwortung von Staaten und multinationalen Konzernen für Ausbeutung und Unterdrückung, nicht abgesehen werden kann. Wie Kelbessa bemerkt, erfordert eine zeitgemäße Umweltethik die Integration unterschiedlicher umweltethischer Ansätze, kultureller Perspektiven und Disziplinen. Daher fordert er den Dialog zwischen indigener Bevölkerung, internationalen Organisationen und Forschern. »All parties should work towards the achievement of the long-term goal of economically, socially, and environmentally sound development« (Kelbessa 2010: 231–232).

—Silvia Donzelli, Berlin

Literatur

Adeh, Ignatius, *Corruption and Environmental Law. The Case of the Niger Delta*, Münster 2010.

Ekwealo, Chigbo Joseph, ›Ndu mmili, ndu azu. Leben und leben lassen: eine afrikanische Umweltethik‹, *Polylog* Nr. 29, 2013, S. 21–36.

Graneß, Anke, *Das Menschliche Minimum*, Frankfurt 2011.

Ikuenobe, Polycarp, »Traditional African Environmental Ethics and Colonial Legacy«, *International Journal of Philosophy and Theology*, Vol. 2, No. 4, 2014, S. 1–21.

Kelbessa, Workineh, *Traditional Oromo Attitudes towards the Environment. An Argument for Environmentally Sound Development*, Addis Ababa 2001.

Kelbessa, Workineh, *Indigenous and Modern Environmental Ethics: A Study of the Indigenous Oromo Environmental Ethic and Modern Issues of Environment and Development*, Washington 2010.

ed. LaMonica, Chris und Omotola, Shola, J., *Horror in Paradise. Frameworks for Understanding the Crises of the Niger Delta Region in Nigeria*, Durham etc. 2014.

Mangena, Fainos, »Discerning Moral Status in the African Environment«, *Phronimon* Vol. 14, Nr. 2, 2013, S. 25–44.

Mähler, Annegret, »Nigeria: A Prime Example of the Resource Curse? Revisiting the Oil-Violence Link in the Niger Delta«, *GIGA Working Papers* Nr. 120, 2010.

Metz, Thaddeus, »Toward an African Moral Theory«, *The Journal of Political Philosophy*, Vol. 15, Nr. 3, 2007, S. 321–341.

Metz, Thaddeus, »An African Theory of Moral Status: A Relational Alternative to Individualism and Holism«, *Ethical Theory and Moral Practice,* Vol. 15 No. 3, 2012, S. 387–402.

Murove, Munyaradzi Felix, »An African Environmental Ethics based on the Conceps of Ukama and Ubuntu«, in: *African Ethics: An Anthology of Comparative and Applied Ethics,* ed. Murove, Munyaradzi Felix, Pietermaritzburg 2009, S. 315–331.

Ogungbemi, Segun, »An African Perspective on the Environmental Crisis«, in: *Environmental Ethics: Readings in Theory and Application,* ed. Pojman, Louis et al, Belmont u. a. 2001, S. 330–337.

Ogungbemi, Segun, »The Conflict in the Niger Delta and National Interest«, in: *Oil Violence in Nigeria: Checkmating its Resurgence in the Niger Delta,* ed. Ojakorotu, Victor and Gilbert, Lysias, Johannesburg 2010, S. 167–180.

Ojomo, P., »Environmental Ethics: An African Understanding«, *The Journal of Pan African Studies,* Vol. 4, No. 3, 2011, S. 101–113.

Oruka, Odera und Juma, Calestous, »Ecophilosophy and the Parental Earth Ethics«, in: *Philosophy, Humanity, and Ecology. Philosophy of Nature and Environmental Ethics,* ed. Oruka, Odera, Nairobi 1994, S. 115–129.

Ramose, Mogobe, »Ecology through Ubuntu«, in: *African Ethics: An Anthology of Comparative and Applied Ethics,* ed. Murove, Munyaradzi Felix, University of Kwazulu-Natal Press, 2009, S. 70–76.

Singh, Jang and Lakhan, V. C., »Business Ethics and the International Trade in Hazardous Wastes«, *Journal of Business Ethics,* Vol. 8, 1989, S. 889–899.

Tangwa, Godfrey, »Some African Reflections on Biomedical and Environmental Ethics«, in: *A Companion to African Philosophy,* ed. Wiredu, Kwasi, Malden 2014, S. 387–395.

Anmerkungen

* Eine ausschöpfende Darstellung der afrikanischen Umweltethik, welche andere wichtige Positionen, wie die *Parental Earth Ethics* von Odera Oruka, berücksichtigen sollte, ist nicht das Ziel dieser Arbeit.

[1] Ähnliche Probleme an der Schnittstelle zwischen wirtschaftlicher Reform, Umweltschäden und sozialer Ungerechtigkeit zeigen sich aktuell im Zusammenhang mit dem Bau des Staudamm Gibe III in Äthiopien.

[2] Solid Urban Waste Management XXI IUPAC Conference, Rome, 6–8 April 2016.

[3] Sieh auch: Jang Singh/ C. Lakhan, ›Business Ethics and the International Trade in Hazardous Wastes‹, *Journal of Business Ethics* Vol. 8, 1989, SS. 889–899.

[4] Metz räumt allerdings ein, dass es indirekte Pflichten gegenüber Pflanzen und Spezies geben kann.

[5] Siehe zum Beispiel Oruka/Juma (1994); Ekwealo (2013).

[6] Siehe auch Ramose (2009).

[7] Zum Zusammenhang zwischen Ölgeschäft und Gewalt im Niger Delta, siehe Mähler (2010).

Von prämoderner Naturliebe zum (trans)nationalen Umweltaktivismus: Der Fall Rumänien

»Der Wald ist der Bruder des Rumänen«.
(Rumänisches Sprichwort)

1 Das Dilemma der interkulturellen Umweltphilosophie

Die folgenden Überlegungen gehen von einer methodischen Frage der interkulturellen Philosophie aus und machen diese am Beispiel der Umweltphilosophie in einer nicht-abendländischen, nämlich der rumänischen Kultur anschaulich. Jede philosophische Reflektion ist ein situiertes Denken, womit hier keinesfalls individuelle Idiosynkrasien, die Hautfarbe, das Geschlecht und der soziale, ethnische und kulturelle Hintergrund als ein unentrinnbares Verhängnis postuliert werden und jegliche Hermeneutik auf eine Aneignung des Anderen zurückgeführt wird. Situiertheit des Denkens heißt: Während der Leib phänomenologisch als Nullpunkt jeglicher Erfahrung betrachtet wird, ist im Denken dagegen kein Rückgriff auf einen Nullpunkt möglich – anders gesagt denkt niemand aus einem *nowhere* für einen anonymen, gleichsam ort- und zeitlosen Adressaten. Auch geht diese Situiertheit mit der Situativität des Denkens einher, so etwa wenn die Verfasserin dieses Beitrags als Wahlwienerin mit rumänischem Hintergrund ihre Überlegungen einer internationalen deutschsprachigen Leserschaft mitteilt. Sosehr das reflektierende Subjekt ein Individuum ist, dessen Denken Anspruch auf Allgemeingültigkeit erhebt, ist jegliches Philosophieren immer schon in eine konkrete Kommunikationssituation eingebettet. Vor allem im interkulturellen Philosophieren ist es nicht belanglos, ob über die eigene Kultur in der Fremde referiert wird oder ob eine andere Kultur für die eigene Kultur (was ich als Innenperspektive bezeichne) oder für eine dritte gedeutet wird (mit dem entsprechenden Sprachwechsel in der Kommunikation und einem anderen Vorwissen). Diese grundsätzliche Unterscheidung zwischen einer Innen- und einer Außenperspektive ist insofern wesentlich, als dem Philosophieren – oder ist das wieder nur die aufklärerische Perspektive des Westens? – eine kritische Funktion zugeschrieben wird. Eine solche Kritik mag zwar in der Ästhetik weniger zum Tragen kommen, aber sie ist zentral in der Ethik mit ihrer impliziten Normativität und nicht zuletzt in der Umweltethik. Diese kritische Stellungnahme wird gewiss deutlich in der Außenperspektive der interkulturellen Philosophie, wenn sich Angehörige einer westlichen Denktradition für eine nicht-westliche interessieren und das Denkmonopol der eigenen Kultur selbstkritisch hinterfragen. In der Innenperspektive jedoch werfen Angehörige einer nicht-westlichen Denkkultur meistens ein positives Licht auf die eigene Tradition, was aber die Möglichkeit einer (Selbst)Kritik prinzipiell schwächt. Mit einem ähnlichen Dilemma konfrontiert sehen sich auch die *public intellectuals,* wenn sie ihr Land im Ausland vertreten. In diese Schwierigkeit brachte mich selbst die Untersuchung der philosophischen Reflektionen über die Natur bzw. Umwelt in Rumänien, wo die bisher wertvollsten Beiträge ethnophilosophisch gefärbt sind und zu ihrer Zeit politische Implikationen im Sinne des Nationalismus hatten.

2 Naturliebe auf Rumänisch

Meine eigene raumzeitliche Situiertheit bedeutet konkret, dass in meiner Schulzeit in Rumänien der 1980er-Jahre bis zum Überdruss die Rede war von der »Symbiose zwischen Mensch und Natur« in der rumänischen Kultur, was im Übrigen im eklatanten Widerspruch zur staatlichen Entwicklungsstrategie stand.

Die Naturliebe gehört wesentlich zum Selbstverständnis der rumänischen Kultur; um dafür zu argumentieren, griff man meistens auf die Volkskultur zurück, vom invariablen Beginn der Lyrik mit »Grünes Blatt« *(frunză verde)* bis hin zu Sprichwörtern wie dem zitierten Motto »Der Wald ist der Bruder des Rumänen« und der in die Landschaft eingebetteten vernakulären Architektur. Alle diese immer noch in der Ethnographie geläufigen Argumente werden durch die Deutungen der Pastoralballade *Miorița* (dt. *Das Schäfchen*) gekrönt. Dieses Gedicht, das in ganz Rumänien in unzähligen Varianten verbreitet ist[1], gilt als einer der grundlegenden »Mythen« der rumänischen Kultur und hat zu diversen Auslegungen rumänischer Historiker, Volkskundler, Dichter und Philosophen seit dem 19. Jahrhundert geführt, vom Fatalismus und der Liebe zum Tod des rumänischen Volks bis zur gleichsam rhythmischen Abwechslung von Hügeln und Tälern als unbewusster stilistischer Matrize der rumänischen Kultur[2]. In dieser Ballade warnt ein Schaf seinen Hirten, dass zwei andere Hirten ihn aus Neid und Gier ermorden wollen. Statt sich zu wehren und schützen, erteilt der junge, unverheiratete Hirte Anleitungen zu seinem Begräbnis und bittet sein Schäfchen, diesen Mord der Schafherde zu verschweigen und ihn als eine Vermählung mit einer fürstlichen »Himmelsbraut« darzustellen, an der die ganze Natur – die Gestirne, die Berge, die Bäume und die Vögel – teilgenommen haben wird.[3] Diese Verklärung eines tragischen Todes als eine kosmische Hochzeit und die gelassene Annahme des Sterbens als Reintegration in den Kreislauf der Natur wurden als Ausdruck einer prächristlichen Weltanschauung gedeutet, die Parallelen mit anderen archaischen Kulturen aufweist.

Andere Ethnologen, aber auch Philosophen haben insbesondere in der Zwischenkriegszeit die (orthodox-)christliche Dimension der rumänischen Volkskultur und ihrer Naturnähe in den Vordergrund gestellt. Typisch dafür ist das 1936–1938 verfasste Buch *Eine rumänische Weltanschauung* von Ovidiu Papadima (1909–1996), der auch während des kommunistischen Regimes einer der führenden Volkskundler Rumäniens war, dessen angeführtes Buch allerdings erst nach der Wende wieder aufgelegt werden durfte (Papadima 1995). Darin betonte Papadima unablässig die »wunderbare Solidarität des Rumänen mit der Natur« und »sein Vertrauen in die Freundschaft und die Hilfe aller Geschöpfe Gottes« (Papadima 1995: 52). Diese »rumänische Weltanschauung« stütze sich auf drei Säulen: Die Harmonie der Welt drücke eine von Gott geschaffene Naturordnung aus, die landwirtschaftliche Arbeit sei kein Eingriff in die Natur, sondern verwirkliche und vervollständige vielmehr die Naturgesetze, und schließlich habe der Mensch durch Barmherzigkeit (im Sinne des Mitgefühls und der Gabe) der Welt alles zurückzugeben, was er im Überfluss hat (Papadima 1995: 51). Auch machten Legenden und Märchen die Kontinuität zwischen dieser Welt und einem analog imaginierten Jenseits anschaulich. Gott werde als ein konkret agierendes Individuum dargestellt, das die Welt mal mithilfe des Teufels schaffe, mal Rat von den kleinen Tieren wie der Biene und dem Igel einhole (Papadima 1995: 27). Die Welt sei gut geordnet, alle Geschöpfe – sogar jene, die von zerstörerischen Absichten bewegt seien –

nähmen an dieser kontinuierlichen Schöpfung teil und die Naturphänomene seien Epiphanien einer göttlich-kosmischen Ordnung, während die Naturkatastrophen die Bedeutung göttlicher Bestrafungen erhielten. Die Korrespondenzen zwischen den beiden Welten und die kosmischen Wirkungen jeder menschlichen Geste erklärten auch die komplizierten magischen Rituale der Volkskultur.

Es wäre fehl am Platz, in dieser Volkskultur philosophische Abstraktionen zu erwarten. Dafür wird der Leser Papadimas mit poetischen Legenden und Bräuchen belohnt, die für den Ethnologen eine »spirituelle Dynamik der Natur (rum. *fire*)« aufzeigen, in dem Sinne dass der Natur Bewusstsein (jedoch keine Seele) zugeschrieben wird (Papadima 1995: 48). In einzelnen Legenden wird die Erde sogar personifiziert. Die zahlreichen kodifizierten Regeln des Volksglaubens kollidieren zum Teil mit der Ordnung des modernen Lebens. Es sei jedoch kein primitiver Aberglaube, der damals gelegentlich zu einem Widerstand gegen die moderne Landwirtschaft führte – so Papadima –, sondern der Respekt für die Umwelt und insbesondere für die Tiere.[4] Von dieser weisen, göttlichen kosmischen Ordnung leitete Papadima auch Vorstellungen über die Sozialordnung ab, wie das Vertrauen der Rumänen in die Berechtigung »natürlicher« Hierarchien und das Misstrauen gegenüber menschlichen (d. h. sozialpolitischen) Reformen. Die Natur gilt nach Papadimas Sicht der rumänischen Volkskultur als rein und heilig, während der Mensch diesen Zustand der Heiligkeit erst durch die Moral erreichen kann. Auch dürfe der Mensch nicht gegen die Naturordnung agieren, sondern müsse diese vielmehr bewahren und verwirklichen. Im Grunde genommen sei die Welt in der rumänischen Folklore »natürlich ästhetisch, weil sie ethisch vorgestellt wird« (Papadima 1995: 66).

Papadimas Auslegung war nicht singulär im Rumänien der 1930er- und 1940er-Jahre. Der fast gleichaltrige Ernest Bernea (1905–1990) sammelte bis 1952 und dann 1965/1966 (nach seiner Entlassung aus dem Gefängnis) ethnologisches Material zur Vorstellung von Raum, Zeit und Kausalität; dieses wurde philosophisch gedeutet in seinem Buch *Rahmen des rumänischen Volksglaubens*, das 1985 – d. h. noch während des Nationalkommunismus – erschienen ist (Bernea 1985).

3 Natur und Sein

Gleichzeitig suchten die Philosophen derselben Generation nach ontologischen Grundlagen der rumänischen Naturauffassung. Der Soziologe, Ökonom und christliche Philosoph Mircea Vulcănescu (1904–1952) nahm sich während des Zweiten Weltkriegs wie andere Intellektuelle der Zeit eine »methodische Distanzierung von Westeuropa« durch eine Kritik »der westlichen Ontologie von Platon bis Nietzsche« vor (Vulcănescu 2005: 979)[5]. Von diesem ehrgeizigen Projekt sind nur Aufzeichnungen zur impliziten Metaphysik der rumänischen Volkskultur aus den Jahren 1943/1944 erhalten geblieben.[6] Vulcănescu, der während des Kriegs als Technokrat in der Regierung von Marschall Ion Antonescu tätig war, starb 1952 in einem kommunistischen Gefängnis. Jedem Volk eigne seiner Ansicht nach eine spezifische Auffassung des Seins. Seine »phänomenologische Skizze« zur rumänischen impliziten Ontologie stellte Vulcănescu selbst in eine Reihe mit den kulturtypologischen Untersuchungen von Lucian Blaga, Ovidiu Papadima, Ernest Bernea u.a. (Vulcănescu 2005: 1010f.). Konkret versuchte Vulcănescu in seiner Ethnophilosophie, eine spezifisch rumänische Auffassung des Individuums, des natürlichen Seins, der Individuation, der Geschlechter, des Sinns, des Schicksals und letztlich Gottes herauszufinden. Die rumänische Metaphysik ist laut Vulcănescu konkret; in ihrem Mittelpunkt stünde weder das Sein, noch die Dinge, sondern das Individuum *(ins*, von lat. *ens)* im Sinne eines »konkreten metaphysisches Substrats« und einer »lebendigen Einheit der Seinsmodi und der mit ihnen verbundenen Ereignisse« (Vulcănescu 2005: 984). Das Individuum habe aber eine Natur (*fire*, von *a fi*, lat. *fieri:* sein). Dasselbe Wort für die Natur *fire* meint sowohl die Umwelt als auch das Wesen eines Dings und den Charakter eines Menschen. Diese Spezifizierung ist umso notwendiger, als das Rumänische ein zweites Wort für ›Sein‹ kennt, *ființă* (aus dem Vulgärlatein *fientia*); nur dieses ist in die philosophische Terminologie eingegangen[7], und zwar gepaart mit *esență* für ›Wesen‹.

Vulcănescus Ontologie ist aber in erster Linie eine der *fire*. In der Auffassung dieser *fire* in der Volkskultur und Alltagssprache stellt Vulcănescu Ähnlichkeiten mit den mediterranen Kulturen fest. Alle basierten auf einer essentialistischen Metaphysik, in dem Sinne dass die Existenz (Da-sein) auf das Was- und Wie-sein, auf die *quidditas* und *ipseitas* zurückgeführt würden – die auf Rumänisch alle als *fire* bezeichnet werden. Diese Metaphysik des Wesens unterscheide sich von der »existentialistischen« Metaphysik der »nördlichen Völker«, wo das, was ist, sich nicht unmittelbar als Wesen zeige, sondern sich zunächst als Präsenz bekunde (Vulcănescu 2005: 985). Die moderne westliche Metaphysik sei vorrangig »existentialistisch«, weil sie – wie am besten der Fall Heidegger zeigt – die Existenz (Da-sein, Anwesenheit) anstelle der Essenz in den Vordergrund stellt (Vulcănescu 2005: 1330 f.).

Die Neigung der rumänischen impliziten Metaphysik zur Konkretheit zeigt sich für Vulcănescu auch in den Bezeichnungen von Welt *(lume,* von lat. *lumen)*, Raum *(loc,* Ort, lat. *locus)* und Zeit *(vreme)*. So bezögen sich *loc* und *vreme* anders als die abstrakten und in den Naturwissenschaften verwendeten Begriffe *spațiu* und *timp* auf die qualitativen Merkmale der erlebten Raumzeitlichkeit, womit Vulcănescu an die Phänomenologie anschließt. Jedenfalls betont er das »Gefühl einer umfassenden universalen Zusammengehörigkeit«: Jedes Ereignis und jede Handlung schallen in der ganzen Welt wie ein Echo wider, »das Leben eines Menschen ist mit dem Schicksal eines Sterns verbunden« und ein Verbrechen »verdunkelt die Sonne und den Mond« (Vulcănescu 2005: 1023). Die *fire* als Gesamtheit der Seienden verlaufe *geordnet,* aber sie sei nicht vollständig determiniert. Hinter jedem Ereignis werde ein Sinn erahnt und dieser werde in Zusammenhang mit wohlwollenden oder bösen Kräften gedeutet. Die Natur sei keine neutrale Welt bloßer Zufälle, sondern sie »ruft und schweigt, zeigt sich und verbirgt sich« (Vulcănescu 2005: 1024). Das Sichtbare kommuniziere mit dem Unsichtbaren und Menschenhandlungen interagierten mit natürlichen Vorgängen wie in einem Dialog. Das Diesseits werde allseitig von einem Jenseits umhüllt, durchdrungen, gefüllt und erfüllt (vervollständigt). Dabei ähnle das Unsichtbare dem Sichtbaren und es befinde sich nicht außerhalb der Welt, sondern nur in einer anderen Raumdimension, aus der es wirksam in unseren Alltag eingreife. Damit verbunden sei auch die Heiligkeit der *fire*. Letztlich löse sich die »positive Wirklichkeit« in einer »dynamischen und animistischen Pluralität« auf (Vulcănescu 2005: 1025).

Zeitlich betrachtet geht – so Vulcănescu – das konkrete Sein im Sinne von *fire* über das Präsens hinaus und bezieht das Vergangene ein. Im Unterschied zur westlichen Ontologie privilegiere die rumänische Spiritualität nicht die Anwesenheit. Weil die Existenz vorzüglich »prädikativ« verstanden werde (Sein als Was- und Wie-sein), bedeute ›existieren‹ nicht unbedingt ›hier und jetzt sein‹, sondern es reiche, irgendwann und irgendwo zu sein. Ebenso sei ein Ereignis keine Tatsache wie in der gegenwartsorientierten Metaphysik des Westens, sondern werde als der Wechsel eines Zustands gedeutet, was die Affizierbarkeit anstelle der Wirksamkeit in den Vordergrund stelle und sich höchstens als Interaktion abspiele. In dieser Welt werde das Werden und Fließen nicht mit der negativen Vergänglichkeit *(trecere)* und Vernichtung assoziiert, sondern als Metamorphose relativiert. Des Weiteren argumentiert Vulcănescu, dass die rumänische Sprache nur schwache Negierungen kenne; das Nicht-Sein sei nie absolut, sondern es schränke nur das Sein ein oder werde von diesem abgeleitet. Auch gebe es keine absolute Unmöglichkeit und das Mögliche habe Vorrang vor dem Aktuellen, wie die zahlreichen Verbalformen für das Hypothetische auf Rumänisch zeigen. Der Vorrang des Möglichen vor dem Aktuellen impliziere, dass es nichts Unheilbares, Irreparables und Unwiderrufliches gibt. Die Existenz werde zu einem Spiel der Möglichkeiten und weil kein Verlust unwiederbringlich sei, werde das Leben mit einer gewissen Sorglosigkeit betrachtet. Dieses tiefe Vertrauen werde von einer besonderen Bedeutung des Entscheidungsakts begleitet: Die Entscheidung *(hotărâre)* negiere nicht andere Möglichkeiten, sondern sie ziehe allein eine Grenze *(hotar)* innerhalb des Reichs der Möglichkeiten, ebenso wie der Imperativ keinen Befehl bedeute, noch den Willen, etwas zu tun, sondern eher ein Einverständnis ausdrücke bzw. die Annahme des Geschehenen.

Diese Ausführungen wurden von Vulcănescu zum Teil noch während des Kriegs veröffentlicht, ohne in dem damaligen Kontext viel Beachtung zu finden, andere bilden Projektentwürfe, die erst in den 1990er-Jahren publiziert werden durften. 1987 wurden jedoch manche seiner Ideen – wenn auch ohne die Erwähnung des »Kriegsverbrechers« Vulcănescu – von seinem Jugendfreund Constantin Noica (1909–1987) in Umlauf gebracht. Damals versinnbildlichte Noica das »wahre« Philosophieren abseits der offiziellen Ideologie und wurde stark auch in literarischen Kreisen rezipiert, zumal er die alte rumänische Sprache und Alltagsausdrücke ontologisch auslegte – darunter auch das Wort *fire*, das laut Noica eine »konkrete Universalität« aufweise (Noica 1987: 34–41, 313 f.). Den Wörterbüchern folgend legte er drei Hauptbedeutungen von Natur *(fire)* fest: als 1. »Natur (als Gesamtheit), Welt, Erde, Ding/Lebewesen *(făptură)*«, 2. »Natur als schöpferische Kraft, die in jedem Ding/Lebewesen anwesend ist« und 3. »Natur im Sinne von Wesen oder Eigenschaft jedes Dings/Lebewesens« (Noica 1987: 36–38). Die erste Bedeutung schließt etwa die unzugänglichen Teile des Universums (die Sterne und fernen Galaxien – nicht aber die Sonne) aus; diese sind jenseits der Natur, ohne übernatürlich zu sein. In diesem ersten Sinne bezeichnet die *fire* »das gesamte lebendige und wahrnehmbare Sein der Welt« (Noica 1987: 37). In ihrer zweiten Bedeutung bezeichnet *fire* das Sein in seinem wirkenden Sein; ausgeschlossen wird jedoch das vom Menschen Bewirkte und Hergestellte, etwa die Artefakte. Im ersten Sinne bezog sich *fire* eher auf das Seiende, im zweiten Sinne auf das Werden und die schöpferische Kraft des Seins. Die dritte Bedeutung vervielfältigt das Sein in verschiedene Seinsweisen der Dinge bzw. Lebewesen. Die *fire* inkludiert die menschliche Natur, den Charakter, das Temperament und die Gewohnheiten des Menschen; im Mittelalter wurde *fire* auch auf das Denken und das Gemüt bezogen, wie immer noch in geläufigen Ausdrücken. *Fire* könne, so Noica, jedenfalls nicht ohne Weiteres mit *natură* gleichgesetzt werden (Noica 1987: 41): Die Natur im Sinne von *natură* sei gleichgültig dem Sein gegenüber, sie sei abstrakter und werde wissenschaftlich thematisiert; die *fire* finde vielmehr keinen Platz in den modernen Naturwissenschaften. Auch habe *natură* eine umfangreichere Extension als die konkrete *fire:* Ein Stillleben könne nur als *natură moartă* bezeichnet werden, während die *fire* nie tot oder erstarrt sein könne. Im Unterschied zur *natura naturata* und *natura naturans* sei die *fire* immer nur produktiv *(naturans)*. Die Natur könne auch die Seinsweise bezeichnen; die *fire* bezöge sich immer auf die Seinsweise der Dinge *und* auf ihr Bestehen. Schließlich sei *natură* beschränkt in ihren schöpferischen Möglichkeiten, während die *fire* eine »unvollendete Genesis« darstelle.

4 Die Umweltethik nach der Wende

Die Erwartungen, dass nach dem Fall des Kommunismus und der Grenzöffnung auch die Umweltethik von den intensiven Bemühungen profitieren werde, aktuelle philosophische Entwicklungen im Westen zur Kenntnis zu nehmen, haben sich nicht erfüllt. Zeitgenössische Ethiker thematisieren weder die eklatanten Widersprüche zwischen der von Ethnophilosophen postulierten Naturliebe der Rumänen und den reellen wirtschaftlichen Praktiken, noch zählt der Umweltschutz zu den Prioritäten der angewandten Ethik in Rumänien. Zwar vollzog auch die rumänische Philosophie nach 1989 eine

»ethische Wende«, jedoch wurden andere Themen der angewandten Ethik als prioritär betrachtet, wie allen voran die Aufarbeitung der kommunistischen Vergangenheit und deontologische Fragen im juridischen Bereich. Auch für die neue feministische Philosophie bilden Umweltfragen im Großen und Ganzen immer noch einen blinden Fleck.[8] Eine Ausnahme bildet Mihaela Miroiu (geb. 1955), die englischsprachige Literatur zum Ökofeminismus referiert (Miroiu 2002). Die von Murray Bookchin angestrebten demokratischen, egalitären und autarken *ecocommunities* ließen sich zwar nach Miroiu auch in Rumänien identifizieren, allerdings in den früheren, mittelalterlichen Dorfstrukturen. Diese wurden zwar in der sog. Transition unmittelbar nach der Wende neubelebt, allerdings nur als Notlösung, um das Überleben der Familien unter den Bedingungen einer noch prekären Demokratie zu gewährleisten. Bookchins Vision der sozialen Ökologie weist die Bukarester Politikwissenschaftlerin letztlich als ein *wishful thinking* im Zeitalter der Globalisierung zurück (Miroiu 2002: 224 f.). Im Mittelpunkt ihrer Theorie steht vielmehr die Angemessenheit (rum. *convenabilitate*, engl. *convenability*) als moralisches Handlungskriterium, wobei das Erzielen des »Passenden« nicht ausschließlich rational gemäß unpersönlichen Grundsätzen und Normen erfolgt, sondern ebenso Empathie erfordert. Die von Mihaela Miroiu konstatierte größere Nähe der Frauen zur Natur betrachtet sie letztlich als eine Chance für die Erweiterung der Ethik über die Menschen hinaus auf die biotische Gemeinschaft (Miroiu 2002: 230).

Schuld an der langen Zeit schleppenden Sensibilisierung für die Umweltprobleme Rumäniens trägt auch die Politik (vgl. Duțu 1999). Sechs Monate nach dem Dezember 1989 gab es bereits fünf politische Formationen, die sich pro forma als »ökologisch« bezeichneten, ohne ein entsprechendes Programm zu vertreten. Eine dieser Parteien hat sogar 1991 den Umweltminister in einem sozialdemokratischen Kabinett gestellt. Ein Jahr später deckten aber die Medien auf, dass manche ihrer prominenten Mitglieder in die Affäre des Imports von Giftmüll aus Deutschland nach Siebenbürgen (hauptsächlich Sibiu) verwickelt waren, was maßgeblich zum Untergang dieser politischen Bewegung beigetragen hat. Auch wurden NGOs gegründet, die aber zu Beginn eine bescheidene Rolle spielten in einem Kontext, in dem die rumänische postkommunistische Gesellschaft noch unerfahren im Umgang mit der Zivilgesellschaft war; so mussten diese NGOs, deren Ziele meistens auf spezifische Probleme beschränkt blieben, sowohl gegen einen noch defizitären juridischen Rahmen als auch gegen das Misstrauen der Behörden und der Bevölkerung und die schlechte Finanzierung kämpfen (Duţu 1999: 220). Von dieser Lage profitierten Touristiker und andere Unternehmer, die die Umweltzerstörung sogar in Biosphärenreservaten wie dem Donaudelta vorantrieben.

Zur spektakulären Reifung der Zivilgesellschaft in Umweltfragen trug später der sog. Skandal Roşia Montană bei, der in Rumänien zum ersten Mal Massenproteste für den Naturschutz auslöste und zu einem Symbol des Widerstands gegen die Privatisierung der natürlichen Ressourcen wurde.[9] Hier ein paar Eckdaten zum Fall Roşia Montană: Das kanadische Unternehmen Rosia Montana Gold Corporation wollte in den Westkarpaten umstrittene Verfahren zum Goldabbau einführen, in einer für ihre Schönheit bekannten Gegend, wo Gold und Silber seit eineinhalb Jahrtausenden, d.h. seit der prärömischen Antike, gewonnen werden. Dieses Importmodell nach kolonialistischer Rezeptur, wie Journalisten kritisierten, wurde vor allem nach 2003 auch von manchen rumänischen Politikern unterstützt, die unter Korruptionsverdacht stehen. Dabei spielten die Medien eine ambivalente Rolle. Der Widerstand gegen das Projekt ging von einer Handvoll Dorfbewohner aus, die sich weigerten, umgesiedelt zu werden, um daraufhin dank dem Engagement von Experten, Naturschutzaktivisten und Volontären zur »größten Bewegung der Zivilgesellschaft in Rumänien nach dem Krieg« zu werden, was in den Medien gerne als Kampf Davids gegen Goliath dargestellt wird (Goţiu 2013: 8). Zunächst verloren in den ersten Jahren die Rechtsanwälte, die die Dorfbewohner pro bono vertraten, die Prozesse gegen die großen Rechtsanwaltskanzleien des kanadischen Konzerns. Der Kampf galt aus mehreren Gründen als aussichtslos: die Unerfahrenheit der NGOs; das schlechte Image des Aktivismus im postkommunistischen Rumänien und das positive Image ausländischer strategischer Investoren in den Medien als einzige Chance für eine nicht funktionierende Wirtschaft; die in Rumänien verbreiteten Verschwörungstheorien, gepaart mit einem gewissen Fatalismus und Misstrauen gegenüber dem eigenen Vermögen; die Angst bis vor einigen Jahren, als »Linke« abgestempelt zu werden; die fehlende Solidarität zwischen verschiedenen gesellschaftlichen Gruppen; das Gefälle zu den Budgets für Werbung und Kommunikation der Kontrahenten; das Klischee von der Priorität der Industrie vor der Landwirtschaft in den Entwicklungsstrategien; die massive Auswanderung als Ausweg für unzufriedene Bürger; und sogar die imaginären Assoziationen des Goldes mit Reichtum (Goțiu 2013: 338f.). Schrittweise begann aber der Widerstand gegen das Bergwerkprojekt zu wachsen: Künstleraktivisten (hauptsächlich MindBomb) wurden kooptiert, die spektakuläre Protestaktionen veranstalteten, medial wirksame Sprüche schufen, Plakate im Ausland (sogar in den Arabischen Emiraten) verbreiteten und ein »Fân Fest«-Kulturfestival in Roșia Montană veranstalteten; damit kam es zu einer tatsächlichen Anwendung des in Rumänien vor der Wende vielbesprochenen »Widerstands durch Kultur« (Goțiu 2013: 430). Die ungarisch-reformierte Kirche weigerte sich lautstark, ihre Grundstücke an den kanadischen Konzern zu verkaufen, der rumänisch-orthodoxe Metropolit Siebenbürgens predigte gegen das »ökologische Verbrechen« des Projekts und sogar die Fußballnationalmannschaft schlug sich auf die Seite der Protestierenden (zit. n. Goțiu 2013: 355). Hinzu kamen Artikel in ausländischen Medien *(The Ecologist)*, Solidarisierungserklärungen von Prince Charles, Naomi Klein und Vanessa Redgrave und sogar der Dokumentarfilm *Rosia Montana. Dorf am Abgrund* des deutschen Regisseurs Fabian Daub, der u.a. beim Human Rights Filmfestival in Wien gezeigt wurde. Gleichzeitig hoben Experten der Rumänischen Akademie der Wissenschaften die Schwächen des Projekts hervor, während Schriftsteller und Historiker das immaterielle Kulturerbe der Region wiederentdeckten, von lokalen Legenden über Goldsucher und historischen Legenden vom dakischen König Decebal bis zum Widerstandskampf während der Revolution 1848/49 und anderen Protestaktionen gegen ausländische Bergbauinvestoren in der zweiten Hälfte des 19. Jahrhunderts. Auch erhielt Roșia Montană den Besuch einer UNESCO-Kommission usw. Die Protestbewegung erreichte 2013 ihren Höhepunkt, als mehrere Zehntausende Menschen in ganz Rumänien auf die Straße gingen. Seither liegt das Projekt auf Eis. 2017 verklagte der kanadische Konzern den rumänischen Staat wegen angeblicher Verzögerungstaktiken vor einem internationalen Schiedsgericht.

Andere rumänische Protestaktionen gegen Umweltverschmutzung haben eher die Aufmerksamkeit internationaler Umweltaktivisten gefunden als die Unterstützung heimischer Philosophen. So zählte Naomi Klein die in Rumänien als »Aufstand von Pungești« bekannten Proteste, die seit 2013 im Osten des Landes gegen Bohrungen für die Schiefergasexploration eines US-amerikanischen Unternehmens stattfanden, zu anderen weltweiten Bewegungen der »Blockadia«, von Kanada bis Australien und von Großbritannien bis zum

Nigerdelta und China (Klein 2015: 361). In einem solchen Kontext erhält die prämoderne »Rückständigkeit« der Naturliebe der Rumänen in puncto Mentalität (und *nicht* der Wirtschaftspraktiken) die Bedeutung einer Avantgarde, und es kommt zu interessanten Allianzen zwischen der konservativen Landbevölkerung und den jungen urbanen NGOs, deren Mitglieder häufig im Westen akademisch ausgebildet wurden. Bilder wie jenes »einer alten Frau, die ein Kopftuch trägt und einen knorrigen Spazierstock in der Hand hält, unter der Überschrift [...]: ›Wenn deine Großmutter rebelliert, weißt du, dass deine Regierung versagt hat‹« (Klein 2015: 367), werden durch internationale Medienagenturen verbreitet und legitimieren in einem neuen Kontext den traditionellen Widerstand konservativer Landgebiete gegen die Moderne.

Es ist zu hoffen, dass solche Fälle wie Roșia Montană oder Pungești nicht nur die öffentliche Meinung, sondern ebenso rumänische Philosophen für Umweltfragen sensibilisieren werden. Zur Zeit gibt es wenige Übersetzungen von zeitgenössischer Umweltphilosophie (Luc Ferrry, Pascal Bruckner, Rainer Paslack), aber die Philosophen haben angefangen, sich an internationalen Forschungsprojekten zur nachhaltigen Ökobildung zu beteiligen (Dumitru, Stoenescu 2013) und sich sogar eigenständig – indem sie über das rumänische Umweltbewusstsein reflektieren – zu positionieren, wie mehrere Studien Adrian Miroius (geb. 1954) zeigen (Miroiu 1996, 1999, 2000). 2013 sprach der Bukarester Philosophieprofessor selbst die Frage nach der Situiertheit des eigenen Denkens an: Zwar stimme er zu, dass in der modernen Gesellschaft allgemeine Grundsätze regieren sollten, jedoch äußerte er sich besorgt über die Tendenz, diese »with similar standards and procedures« auch privat Oberhand gewinnen zu lassen und begründete diese Besorgnis mit seinem Aufwachsen in einem autoritären Regime (Miroiu 2013: 133). Stattdessen verlangt Miroiu, in der Umweltethik Platz auch für die Tugenden der Fürsorge, des Teilens und des Austausches *(sharing)*, der Liebe, Einfühlung und Menschlichkeit zu schaffen. Damit kritisiert er zugleich die instrumentelle Betrachtung der Umwelt *und* der Individuen im Kommunismus als Ressource, die zu erobern und rücksichtslos für kollektive Zwecke einzusetzen sei: »So, people's daily life and nature were in the same boat: fellow subjects to the same source of domination« (Miroiu 2013: 135). Das moralische Verhältnis zur Natur müsste vielmehr auf eine persönliche Grundlage gestellt werden, und affektive Werte statt Pflicht und Gerechtigkeit sollten in den Mittelpunkt rücken.

Des Weiteren unterscheidet Miroiu innerhalb der Umweltethik zwischen der sog. »environmental philosophy and environmental ethics« und den »less recent, but neglected, rejected, undervalued or even lost worldviews, among them the native American or the aboriginal Australian outlooks« (Miroiu 2013: 135). Seine eigene Denkposition verortet er allerdings in einem dritten Lager, das von der byzantinischen Naturphilosophie geprägt ist. So weist Miroiu zunächst die in der *environmental ethics* geläufigen Dichotomien zurück (etwa in der Fragestellung, ob die Menschen zur Natur gehören oder nicht oder in der Auffassung der Wildnis als vollständig frei vom menschlichen Eingriff). Das theologische Paradigma der Leibwerdung Christi habe nicht nur den absoluten Gegensatz zwischen Transzendenz und Immanenz, Geist und Körper im Menschen aufgehoben, sondern auch der Schöpfung eine neue Würde verliehen, die der Unterdrückung der Natur jegliche Berechtigung entzieht: »God's humility and human haughtiness do not match very well« (Miroiu 2013: 137). Miroiu zeigt sich erstaunt, dass die ostkirchliche Auffassung der Natur in der *environmental philosophy*[10] vernachlässigt wurde, »for orthodoxy is an essential part of the European tradition, not an exotic world-view« (Miroiu 2013: 138). In der Sicht der Ostkirche habe aber die Fleischwerdung Christi Folgen für die ganze Natur im Sinne der Heiligung und Rettung der ganzen Schöpfung. Miroius Schlussfolgerung, dass es aus dieser Perspektive keinen Platz für die Sündigkeit des Fleisches geben könne, mag zwar Theologen irritieren, dass aber die Natur rein sei und Sünde nicht kenne, dürfte weniger problematisch sein. Das ökologische Engagement der Orthodoxie ist im Übrigen nicht nur erst seit dem sog. »grünen Patriarchen« Bartholomäus bekannt. Dieser Bezug auf die Theologie der Ostkirche dient jedenfalls Miroiu dazu, Baird J. Callicotts Verallgemeinerung, dass in der jüdisch-christlichen Überlieferung Gott (bloß) transzendent ist, zu verwerfen.[11] Dualismen und die sowohl in der Umweltethik als auch in der feministischen Philosophie kritisierten Hierarchien (Natur, Leib, Affektivität, das Weibliche und die Subjektivität als der Kultur, dem Geist, dem Männlichen und der Objektivität unterlegen) seien der Orthodoxie fremd, so Miroiu: »the Christian orthodox religion provides a basis for a genuine environmental ethics. Natural items have intrinsic value« (Miroiu 2013: 139). Die Polemik zwischen Callicott und Rolston in Bezug auf den intrinsischen oder objektiven Wert der Natur wird von Miroiu ignoriert und er bezieht sich undifferenziert auf beide US-amerikanischen

Umweltethiker. Diese Unterscheidung mag aber für Miroiu insofern irrelevant sein, als der Kern seiner Argumentation besagt, dass a) sich der Wert der Natur vom Wert der Menschen ableitet und dass b) dies keine subjektive Auffassung des Wertes der Natur impliziert. Ganz im Gegenteil basiert für ihn der intrinsische/objektive Wert der Natur auf dem *Leibwerden* Christi zur *Rettung* der ganzen Schöpfung. Dabei handelt es sich wohlgemerkt nicht um Grundsätze, sondern um allgemeine Bedingungen einer Umweltethik, die Chancen hätte, von den Rumänen akzeptiert zu werden. Zusammenfassend müsste eine für Rumänien geeignete Umweltethik für Adrian Miroiu 1. sowohl Pflichten und Rechte als auch die angeführten affektiven *green virtues* berücksichtigen, damit 2. das umweltbewusste Verhalten dem Bürger nicht »from a public position« und mit »oppressive results« erzwungen werde; schließlich würde 3. eine solche Theorie den intrinsischen Wert der Natur postulieren (Miroiu 2013: 139f.). Damit ist keine *rumänische* Umweltethik intendiert, die einem vermeintlich zeitlosen und abstrakten Volksgeist entspräche. Miroius Analyse geht vielmehr vom heutigen kulturellen Kontext in Rumänien aus und betrifft weniger den Inhalt einer spezifischen umweltethischen Theorie als strukturelle Bedingungen und Einschränkungen (sog. *structural* bzw. *epistemological constraints*), die erklären, warum manche ethischen Theorien akzeptabler als andere erscheinen.

Fazit: Angesichts der Vorbelastung der rumänischen Ethnophilosophie und der Dringlichkeit der philosophischen Reflexion über die aktuellen Wirtschaftspraktiken und über einen auch in Rumänien angekommenen Lebensstil, der alles andere als nachhaltig mit den natürlichen Ressourcen umgeht, ist eine Rezeption der *environmental philosophy* durchaus zu begrüßen. Das Beispiel Adrian Miroius zeigt, dass die Lektüre westlicher Philosophie eine zweifache Kritik einbeziehen kann: des modernen Denkens und der selbst kritischen zeitgenössischen Umweltphilosophie. Eine Suche nach Allianzen in anderen Kulturen fand sich allerdings bisher nicht, vielmehr wurde verlangt, das Verständnis des europäischen Denkens zu erweitern, um andere bisher marginale philosophische und theologische Traditionen einzubeziehen. Meine zu Beginn formulierte Forderung an die interkulturellen Philosophen, die eigene Denksituiertheit mitzureflektieren, wird hier nebenbei vollzogen. Nicht zuletzt treffen sich die angeführten Vertreter der zeitgenössischen rumänischen Philosophie trotz ihrer völlig anderen Denkgenealogie als die ältere Generation der Ethnophilosophen doch insofern mit ihnen, als auch sie theoretische Antworten suchen, die sich sowohl für die eigene Kultur eignen als auch zur internationalen Diskussion beitragen.

—*Mădălina Diaconu, Universität Wien, Österreich*

Literatur

Bernea, Ernest, *Cadre ale gîndirii populare românești*, Bukarest, 1985.

Blaga, Lucian, *Trilogia culturii. Orizont și stil. Spațiul mioritic. Geneza metaforei și sensul culturii*, Bukarest, 1944.

Bran, Paul, *Dimensiunea economică a impactului de mediu. Studiu de caz: Roșia Montană*, Bukarest, 2004.

Cocean, Pompei, *Roșia Montană in universal history*, Cluj-Napoca, 2012.

Daub, Fabian, *Rosia Montana. Dorf am Abgrund*, Dokumentarfilm D/RO, 2012, 78′.

Dumitru, Daniela, Constantin Stoenescu, »The Ethos of Sustainable Development and the Educational System: A Research Report regarding Romania«, in: ed. Mureșan, Valentin and Majima, Shunzo, *Applied Ethics. Perspectives from Romania*, Kitaku Sapporo, 2013, pp. 144–175.

Duțu, Mircea, »Despre ecologismul românesc«, in *Ecologie. Filosofia naturală a vieții*, Bukarest, 1999, pp. 219–232.

Eliade, Mircea, »Das weissagende Lämmchen«, in *Von Zalmoxis zu Dschingis-Khan*, Köln-Lövenich 1982, pp. 235–267.

ed. Fochi, Adrian, *Miorița*, Bukarest, 1964.

Frunză, Mihaela, *Tematizări în eticile aplicate. Perspective feministe*, Cluj-Napoca, 2009.

Goțiu, Mihai, *Afacerea Roșia Montană*, Cluj-Napoca, 2013.

Klein, Naomi, *Die Entscheidung Kapitalismus vs. Klima*, Frankfurt am Main, [2]2015.

Miroiu, Adrian, »Ethical Attitudes toward the Environment: The Romanian Case«, in: ed. Mureșan, Valentin, and Majima, Shunzo, *Applied Ethics. Perspectives from Romania*, Kitaku Sapporo, 2013, pp. 131–143.

Miroiu, Adrian, »Global Warming and Moral Theorizing«, in *Theoria*, Vol. 11, 1996, pp. 61–81.

Miroiu, Adrian, »The Conceptual/Normative Distinction in Environmental Ethics«, *Revue Roumaine de Philosophie*, Vol. 42–43, 1999, pp. 133–145.

Miroiu, Adrian, »Two Approaches to Intrinsic Value«, *Revue Roumaine de Philosophie*, Vol. 44, 2000, pp. 367–376.

Miroiu, Mihaela, »Ecofeminismul si problema ›naturalizării‹«, in *Convenio. Despre natură, femei și morală*, Iași, [2]2002, pp. 197–233.

Noica, Constantin, *Cuvînt împreună despre rostirea românească*, Bukarest, 1987.

Papadima, Ovidiu, *O viziune românească a lumii*, Bukarest, 1995.

Pascaru, Mihai, *Globalizare românească. Impactul comunitar al proiectului Roşia Montană Gold Corporation*, Floreşti-Cluj, 2013.
Vulcănescu, Mircea, *Opere I. Dimensiunea românească a existenţei*, Bukarest, 2005.

Anmerkungen

1 Adrian Fochi veröffentlichte über 700 Varianten dieser Ballade in *Mioriţa* (Bukarest, 1964).
2 Lucian Blaga, *Orizont şi stil* (1935) und *Spaţiul mioritic* (1936), in Blaga (1944: 5–155 bzw. 157–336). Vgl. auch die Zusammenfassung dieser Deutungen bis in die 1960er-Jahre in Eliade (1982).
3 »Sag ihnen frei: / Daß ich vermählet sei / mit einer Fürstin traut, / mit einer Himmelsbraut; / als es die Hochzeit gab, / fiel hell ein Stern herab; / Sonne und Mondenglanz / hielten den Hochzeitskranz, / Espe war, Tanne war / unter der Gästeschar; / Berge die Priester war'n / Spielleut die Vogelschar'n / – mochten wohl tausend sein – / Sterne: der Fackelschein« (Übers. v. Alfred Margul Sperber, zit. n. Eliade 1982: 239).
4 Wenn etwa zwei Wochen nach dem Epiphanie-Fest, wenn von den Priestern die Gewässer geweiht werden, keine Wäsche gewaschen wird, um das Wasser nicht zu beschmutzen (Papadima 1995: 97). Ebenso haben damals Landwirte den Gebrauch von Insektiziden abgelehnt, weil die ganze Natur heilig ist.
5 Mircea Vulcănescu, »Despre spiritul românesc«, in Vulcănescu (2005: 974–979).
6 Vulcănescu, »Existenţa concretă în metafizica românească« (verfasst 1943/44, veröffentlicht erst nach der Wende) und »Dimensiunea românească a existenţei. Schiţă fenomenologică« (1943), in Vulcănescu (2005: 980–1006 bzw. 1007–1062). Letztere Schrift wurde Emil Cioran gewidmet.
7 So wurde Heideggers *Sein und Zeit* als *Fiinţă şi timp* übersetzt.
8 Typisch dafür ist Mihaela Frunzăs Abhandlung *Tematizări în eticile aplicate. Perspective feministe*, in der Umweltfragen eklatant fehlen und allein die Tierrechte beiläufig erwähnt werden – und dies auch nur im Zusammenhang mit den Rechten von Ungeborenen (Frunză 2009: 149).
9 Zu der umfassenden Bibliographie rund um Roşia Montană zählen historische Studien (Cocean 2012), Wirtschaftsanalysen (Bran 2004) und soziologische Impaktstudien (Pascaru 2013). Die Geschichte dieses Falls findet sich genau dokumentiert auf über 500 Seiten in *Afacerea Roşia Montană* (2013) des Journalisten Mihai Goţiu.
10 Damit ist die nordamerikanische Umweltphilosophie gemeint; deutsche Beiträge zu Umweltfragen wurden in der rumänischen Philosophie bisher kaum rezipiert.
11 Jesus »is not intangible and far, too far from us«, »His dwelling is in our innermost nature« (Miroiu 2013: 138).

Planetarische Integrität – Was Umweltethiker und interkulturell interessierte Philosophen voneinander lernen können

1 Einleitung

Als am 22. März 2017 das neuseeländische Parlament dem Whanganui Fluss Rechte zusprach, war das eine kleine Sensation. Es war nicht nur der erste Fluss auf der Erde, dem als juristische Person Rechte zugesprochen wurden, sondern es war auch die Beilegung eines über 170 Jahre andauernden Rechtsstreites zwischen der neuseeländischen Regierung und dem Maori Stamm. Die Maori hatten seit Mitte des 19. Jahrhunderts für die Beibehaltung bzw. Anerkennung ihrer Kultur gestritten. Adrian Rurahwe, ein Vertreter der Maori, kommentierte die Zuerkennung des Personenstatus für den Whanganui Fluss mit den folgenden Worten: »It's not that we've changed our worldview, but people are catching up to seeing things the way that we see them« (Davison 2017).

Was war geschehen, dass es im Jahre 2017 plötzlich möglich wurde, das holistische Weltbild der Ur-Einwohner anzuerkennen? Seit 1831 die ersten Europäer in Neuseeland ankamen, hatte die koloniale Matrix, die Flüsse als Sachen betrachtete, die Oberhand behalten. Erst in den vergangenen Jahrzehnten, war es möglich geworden, dem Fluss zunächst den Namen der Eingeborenen »Whanganui« zurückzugeben und ihn innerhalb eines Nationalparks unter besonderen Schutz zu stellen. Die rechtliche Anerkennung bzw. Nachbildung der Bedeutung, die der Fluss innerhalb der eingeborenen Kultur hat, ist dann im März 2017 vollzogen wurden und lehnt sich an das folgende Sprichwort der Maori an: »1. Ko te Awa te mata puna o te ora – The River is the source of spiritual and physical sustenance. 2. E rere kau mai te Awa nui mai te Kahui Maunga ki Tangaroa – The great River flows from the mountains to the sea. 3. Ko au te Awa ko te Awa ko au – I am the River and the River is me. 4. Nga manga iti, nga manga nui e honohono kau ana, ka tupu hei Awa Tupua – The small and the large streams that flow into one another and form one River« (Bosselmann 2017: 165).

Die neuerliche Anerkennung des Whanganui Flusses als Träger subjektiver Rechte ist Ausdruck der *interkulturellen Situation*, in der wir leben. Der Umstand, dass der normative Status, der dem Fluss innerhalb der Kultur der Maori zukommt, in geltendes Recht übersetzt wird, zeigt an, dass seitens der Juristen keinerlei Notwendigkeit besteht, die rechtsstaatliche »Rationalität« gegen das Weltbild der neuseeländischen Ur-Einwohner auszuspielen. Demgegenüber ist es eine geradezu als hoffnungsvoll zu bezeichnende Hinwendung zu den prä-kolonialen Konzepten der Maori. Sie verfügten über alternative Ontologien und differente Wert-Sprachen *(languages of evaluation)*. Oft werden dabei unabdingbare materielle und spirituelle Grundlagen des Lebens benannt, die sich innerhalb moderner Nationalstaaten in ein Recht auf Leben und Gesundheit übersetzen lassen. Die Landrechte indigener Völker auf angestammte Territorien werden dabei oft auf die Bedeutung dieser Orte innerhalb spiritueller Praktiken der Eingeborenen zurückgeführt. Auf diese Weise können sich eigenständige Konzepte der Umweltgerechtigkeit ergeben, die zwischen strategischen Zielen und Aspekten der Weltanschauung vermitteln. »Die Verwendung verschiedener Wert-Sprachen stellt nicht nur eine Strategie dar, um Entschädigung für externe Effekte zu erlangen, sondern gründet in tief verwurzelten kulturellen Werten« (Martinez-Alier 2016: 228). Die je konkrete Verteidigung von Wäldern, Flüssen und bestimmten Landstrichen ist also untrennbar verbunden mit der Verteidigung bestimmter Lebensformen, deren Differenz zum europäisch-westlichen Denken – insbesondere der Sprache der Ökonomie – im Schulterschluss von Indigenen und Wachstumskritikern betont wird. Es wird daher auf die Schwierigkeit hingewiesen, einen Kompromiss zwischen verschiedenen Ontologien und damit verbunden normativen Urteilen zu finden: »Solche Wertsprachen lassen sich oft nicht ineinander übersetzen« (Martinez-Alier 2016: 228).

Im Medium das Rechts ist es scheinbar doch möglich, differierende Wertsprachen »aufzuheben« und als lokale ethische Rechtfertigungen einer allgemeineren Terminologie der Jurisdiktion zu betrachten. Damit ist implizit ein Anliegen der interkulturellen Philosophie verbunden, nämlich die Anerkennung verschiedener Weltanschauungen als gleichberechtigte Sichtweisen auf die Welt. Gerade dieses Merkmal erweist sich als Schlüssel zum Verständnis des gegenwärtigen *Rights of Nature*-Diskurses. Diese Anerkennung findet vielleicht im Fall der Verfassung von Ecuador und den Rechten der Natur, die sich unter anderem auf die Naturvorstellung *Pachamama*[1] der Andenvölker beziehen, ihren prominentesten Ausdruck. Auch für das Verständnis der neuerlichen Rechte des Whanganui Flusses ist sie einschlägig. In einem weiteren Fall der Zuerkennung der Rechte für zwei Gletscher im Himalaya Gebirge in Indien spielt dieses Motiv eine Rolle (vgl. Knauß 2018).

Ziel dieses Textes ist es, Gemeinsamkeiten in den Fällen der Rechte der Natur seit 2008 in eine abstraktere philosophische Terminologie zu übersetzen. Anhand der Debatte der Rechte der Natur lässt sich ein synergetisches Zusammenspiel zwischen interkultureller Philosophie und Umweltethik skizzieren. Der Zielpunkt meiner Argumentation wird das Konzept der *planetarischen Integrität* sein, ein deskriptiv-normatives Gefüge, das einerseits dazu in der Lage ist, eine polyzentrische Welt- und Wissensordnung abzubilden, andererseits drängende Fragen der Gegenwart, wie die menschengemachte Erderwärmung zu thematisieren. *Planetarische Integrität,* so meine Auffassung, kann einerseits eine synthetische Einheit abbilden, die sich mit dem dekolonialen Denken[2] oder der interkulturellen Philosophie begründen lässt. Andererseits ermöglicht uns das Konzept, insbesondere die Interaktion zwischen den Menschen und ihrer planetaren Umwelt zu begreifen.

2 Zur interkulturellen Philosophie

Bei der Beschäftigung mit interkulturellen Fragestellungen in der Philosophie sind zwei Aspekte von besonderer Bedeutung. Terminologisch lassen sich diese beiden Elemente als *Exteriorität* und *Universalität* bezeichnen, eine mögliche Synthese der beiden als *Integrität.*

Culture matters! – Die Bedingung der Exteriorität

Um überhaupt sinnvoll von Interkulturalität sprechen zu können, müssen wir eine inhaltliche Verschiedenheit der Kulturen annehmen, die ich als *Exteriorität* bezeichne. Die Verschiedenheit der Kulturen bedeutet keine selbstgenügsame Authentizität, da wir längst wissen, dass Kulturen keine monolithischen Wesenheiten, sondern gemachte

Mischverhältnisse sind, die sich gegenseitig beeinflussen, ineinander übergehen und sich manchmal bis zu Unkenntlichkeit überlappen. Trotz des Hybriden, das mir nicht die Ausnahme, sondern die Regel der Kulturen zu sein scheint, benötigen wir ein Zugeständnis, dass sich Kultur A von Kultur B unterscheiden kann, damit die Konstruktion der interkulturellen Philosophie Sinn macht. Es liegt daher nahe, Kultur phänomenologisch mit einem Moment der *Fremdheit* zu identifizieren, das sich im Kontakt mit anderen zeigt. Es ist zwar nicht der Fall, dass wir im interkulturellen Kontakt permanent die Erfahrung der Fremdheit machen, auf Unverständnis oder sogar Ablehnung stoßen, dennoch wäre es falsch, diese Erfahrung der Andersartigkeit von Handlungs- und Sinnzusammenhängen zu leugnen.

Ein wichtiger Meilenstein bei der Beschreibung des Phänomens des Fremden ist Edmund Husserl. Für Husserl bedeutet der Andere den »original Unzugänglichen« (Husserl 1963: 144). Obwohl Husserl von der originalen Unzugänglichkeit spricht, ist diese Verschiedenheit nicht im Sinne einer absoluten metaphysischen Differenz bzw. einer unüberbrückbaren wesensmäßigen Verschiedenheit zu verstehen, sondern zunächst empirisch als Erfahrung des Fremden zu begreifen. Dies wird deutlich, wenn man an Husserls Definition des Subjekts als »Ichpol« denkt, eine im Grunde inhaltsleere, funktionale Bestimmung des Ich als Knotenpunkt unserer Erfahrungen. Ohne die Erfahrungen, die wir machen, sind wir gleichsam nichts. Subjekte sind für Husserl also keine ursprünglichen Wesenheiten, sondern lediglich Konglomerate aus Erlebnissen, Eindrücken, Wünschen und Hoffnungen. Was den anderen unterscheidet, ist folglich »bloß« seine differente Lebenserfahrung, die ich selbst durchaus hätte genauso machen können, die ich faktisch aber einfach anders erlebt habe. »Ich kann die Erfahrung des Anderen also nachvollziehen, sie wird aber nie deckungsgleich mit der Erfahrung des Anderen sein, weil der Andere, während ich seine eben gemachte Erfahrung nachvollziehe, seinerseits bereits neue Erfahrungen macht. Als originale Erfahrung bleibt die Erfahrung des Anderen unzugänglich« (Weidtmann 2016: 77).

Vor allem in der französischen Phänomenologie ist das Thema der Exteriorität bearbeitet worden. Bei Maurice Merleau-Ponty erfolgt der Zugang zum Phänomen des Anderen dabei über Intersubjektivität und Leiblichkeit. Zum einem geht er davon aus, dass Intersubjektivität der Subjektivität vorausgeht, zum anderen nimmt er an, dass wir die Welt und den Anderen nur vermittels unseres Leibes erfahren. Der eigene Leibe findet im fremden »so etwas wie eine wunderbare Fortsetzung seiner eigenen Intentionen, eine vertraute Weise des Umgangs mit der Welt«, so dass »der fremde Leib und der meinige ein einziges Ganzes, zwei Seiten eines einzigen Phänomens« bilden (Merleau-Ponty 1966: 405). Da Intersubjektivität bei Merleau-Ponty in erster Linie körperlich konnotiert ist, wird die Begegnung mit dem Anderen auch als »Zwischenleiblichkeit« bezeichnet (Merleau-Ponty 2004: 185). Die Praxis der Begegnung ist ein »gemeinsames Tun«, »deren Schöpfer keiner von uns beiden ist« (Merleau-Ponty 1966: 406). Erst im Nachhinein wird die Erfahrung »zu einer Episode meiner privaten Geschichte« (Merleau-Ponty 1966: 406). Als Beispiel führt Merleau-Ponty das Gespräch an, das keineswegs ein bloßes Austauschen von Argumenten darstelle, sondern von vielfältigen Momenten der Interaktion gekennzeichnet sei. So bestimme der Gesprächsverlauf wesentlich die Gedanken und sprachlichen Äußerungen der Teilnehmer.

Bei Emmanuel Lévinas erfolgt die Thematisierung des Anderen weder als *alter ego* (Husserl) noch als Aspekt leiblich erfahrener Intersubjektivität (Merleau-Ponty). Lévinas rückt das Andere als Anderes in den Mittelpunkt seiner Überlegungen. Er schließt sich dabei zunächst Heideggers Auffassung an, das menschliche Dasein sei wesentlich »In-der-Welt-sein«. »Welt« ist kein objektiver Gegenstand, zu dem man sich verhalten könne, sondern vielmehr ein Lebens- und Interpretationszusammenhang, auf den wir uns immer schon eingelassen haben. Lévinas deutet diesen Zusammenhang daher als »Totalität«[3], um den umfassenden Charakter unserer ursprünglichen Welterfahrung anzuzeigen. Das »Antlitz«[4] des Anderen sei nun in der Lage, diese Totalität aufzubrechen. Die Andersheit des Anderen sei »jenseits des Seins« zu verorten und stelle daher die »Spur der Unendlichkeit« dar (Lévinas 2002: 35:65). »Das Antlitz ist für Lévinas also so etwas wie ein Fenster der Welt, durch das wir die jenseits der Welt liegende Unendlichkeit erfahren können« (Weidtmann 2016: 92). Das Gewahrwerden des Fremdpsychischen deutet an, dass die Welt, die wir selbst wahrnehmen, möglicherweise anders betrachtet wird. Wie genau wissen wir nicht, deshalb hat der Blick des Anderen stets etwas Rätselhaftes und Geheimnisvolles.

Der deutsche Phänomenologe Bernhard Waldenfels spricht ebenfalls vom Fremden und bezeichnet es als »Stachel« im Fleische des Eigenen (Waldenfels 1990: 93). Er schließt dabei an Husserl an: Fremderfahrung ist für Waldenfels »eine Form der Erfahrung, nur eben in der paradoxen Form einer originären Unzugänglichkeit, einer abwesenden Anwesenheit« (Waldenfels 1997: 30). Waldenfels weist dabei eine doppelte Erfahrungsdimension der Fremdheit aus: »Nehmen wir das Fremde dagegen als etwas, das nicht dingfest zu machen ist, nehmen wir es als etwas, das uns heimsucht, indem es uns beunruhigt, verlockt, erschreckt, indem es unsere Erwartungen übersteigt und sich unserem Zugriff entzieht, so bedeutet dies, daß die Erfahrung *des* Fremden immer wieder auf unsere eigene Erfahrung zurückschlägt und in ein *Fremdwerden der Erfahrung* übergeht« (Waldenfels 2006: 7 f.). Zum einen ist Fremdheit ein möglicher Aspekt unserer Erfahrung, zum anderen ist das Fremde die Bedingung *sine qua non* der Erfahrung selbst. Dies lässt sich leicht veranschaulichen: Selbst beim wiederholten Erfahren an sich bekannter Gegenstände können immer wieder neue Eigenschaften derselben entdeckt werden. Das Fremde bildet somit Grund dafür, dass wir überhaupt Neues begreifen können.

Waldenfels nimmt weiterhin *Steigerungsgrade des Fremdseins* an (Waldenfels 1997: 35 ff.). Als »alltägliche Fremdheit« bezeichnet er den Aspekt, dass wir beinahe täglich Menschen und Dingen begegnen, die wir noch nie zuvor gesehen haben. Auch geht er von »struktureller Fremdheit« aus, wie sie bei Fremdsprachen vorliegt: »Die Fremdsprache gehört nicht in die eigene Lebensordnung, bleibt aber doch grundsätzlich in sie übersetzbar, d. h. sie gehört einer anderen Ordnung an, die zwar anderen Ordnungsmustern folgt, darin dass sie ordnet aber vertraut bleibt« (Weidtmann 2013: 95). Als »radikale Fremdheit« bezeichnet Waldenfels schließlich das »Außerordentliche«, also Erlebnisse, die uns nicht strukturell begegnen und neben der Ordnung stehen und uns zur Modifikation der Ordnung selbst anregen. »Fremdheit in ihrer radikalen Form besagt, daß das Selbst auf gewisse Weise *außer sich selbst* ist und daß jede Ordnung von Schatten des Außer-ordentlichen umgeben ist. Solange man sich dieser Einsicht verschließt, bleibt man einer *relativen* Fremdheit verhaftet, einer bloßen Fremdheit für uns, die einem vorläufigen Stand der Aneignung entspricht« (Waldenfels 2006: 116).

Es hat sich gezeigt, dass *Exteriorität* ein zentrales Thema der Phänomenologie im Laufe des 20. Jahrhunderts gewesen ist. Merleau-Ponty expliziert Andersheit zunächst auf der Ebene des Leibes. Lévinas behält diese leibliche Konnotation bei, betont ausgehend vom Antlitz des Anderen aber die ethisch-ontologische Fundamentaldimension von Andersheit. Das Andere ist hiernach Bedingung *sine qua non* von Erfahrung und die ursprüngliche Quelle der Verantwortung. Über Waldenfels' Einführung der Alterität als Fremdheit in die deutsche akademische Diskussion ergibt sich auch ein Anknüpfungspunkt für die interkulturelle Philosophie. Gerade weil Fremdheit in den verschiedenen Graduierungen von alltäglicher Fremdheit, struktureller Fremdheit und bisweilen auch radikaler Fremdheit ein konstitutiver Bestandteil menschlichen Lebens ist, liegt es nahe, *Exteriorität* als Merkmal von Kulturen ernst zu nehmen. Kulturen stellen dabei kollektive Sinn- und Handlungszusammenhänge für Individuen dar. Sie werden aber auch als Grund der Fremdheit von Interpretationen und Praktiken betrachtet. Dabei darf das Diktum *culture matters* aber nicht dauerhaft in zwei Extreme verfallen: *Inkommensurabiltität* und *identitätsleere Einheitskultur.* Kulturen existieren im Plural, die Vernunftkultur oder die identitätsneutrale liberale Einheitskultur kann und darf es m. E. nicht geben. Kulturen sind stets konkret, raumzeitlich situiert und mit besonderen Erfahrungen verbunden, die in spezifischen Sichtweisen und Präferenzen münden. Trotz dieser Verschiedenheit der Kulturen kann es durchaus Gemeinsames geben. Menschen verschiedener Kulturen haben bei genauerem Hinsehen sogar sehr viele Gemeinsamkeiten, die zunächst aus ihrer faktischen Begegnung herrühren und die im Weiteren auch Hoffnung auf eine Einigung in wesentlichen normativen Grundüberzeugen machen. Hiermit ist bereits das zweite Element angesprochen, das ich für unabdingbar halte.

Einigung ist möglich, die Bedingung der Universalität

Warum soll es denn notwendig sein, neben der Verschiedenheit der Kulturen überhaupt an einer gemeinsamen Dimension festzuhalten? Die Berufung auf Universalität scheint mir notwendig, um ein Missverständnis zu vermeiden, das ich als *Authentizitätsfalle* bezeichne. Die suggestive Behauptung, etwa der lateinamerikanischen Denker der *Dekolonialität* Enrique Dussel, Anibal Quijano und Walter Mignolo als Sprachrohr der Opfer aufzutreten, kokettiert mit dem Anschein von *wahrhaftiger Repräsentation.* Die Berufung auf *Exteriorität* tappt in die *Authentizitätsfalle,* wenn sie Kulturen aufgrund ihrer zugestandenen inhaltlichen Verschiedenheit als inkommensurabel darstellt. Hier wird man immer die Frage stellen können und müssen, ob ein Sprecher tatsächlich im Sinne der Anderen spricht oder nicht.

Ein Vorteil der Redeweise des Sprechens für Andere, bei aller Unmöglichkeit, *authentisch zu sprechen*, ist es, wenigstens mit offenen Karten zu spielen. Wenn Kant mit der Stimme der reinen Vernunft, Rawls mit der Stimme des *Overlapping Consensus* oder Habermas für den herrschaftsfreien Diskurs spricht, dann wird die Repräsentation einfach unsichtbar gemacht. Methodisch wird das dann als »view from nowhere«, »the God's eye view« oder »objektive Perspektive« der Wissenschaft bezeichnet. Dies ist meines Erachtens zu Recht als Hybris der Moderne bezeichnet worden.

Zusammenfassend lässt sich sagen, dass die Verschiedenheit menschlicher Kulturen unweigerlich eine inhaltliche Dimension bedeutet, die ich mit dem Terminus *Exteriorität* gekennzeichnet habe. In einem gewissen produktiven Spannungsverhältnis zu dieser Andersartigkeit steht die Gemeinsamkeit der Kulturen, die als *Universalität* zu bezeichnen ist. Meinem Verständnis nach lassen sich Universalität und Exteriorität nicht aufeinander reduzieren. Die Hoffnung des Schleiers des Nichtwissens ist die Gleichheit, seine Gefahr ist die Verdeckung der je konkreten Identitäten, die menschliche Subjektivität unweigerlich bedeutet. Die Hoffnung der *Exteriorität* ist, diesen Schleier zu lüften, ihre Gefahr ist die Konzeption eines Außen, zu dem es kein Innen (mehr) gibt. Um dem produktiven Spannungsverhältnis von Exteriorität und Universalität und der Manifestation von Zwischenergebnissen, die dieses Zusammenspiel zeitigt, einen Namen zu geben, führe ich den Begriff der *Integrität* ein.

3 Planetarische Integrität

Um das produktive Spannungsverhältnis von Universalität und Exteriorität zu denken, hat Immanuel Wallerstein von einem »universalen Universalismus« gesprochen, der für ihn bedeutet:

> eine ständige Spannung zu akzeptieren, nämlich zwischen dem Bedürfnis, unsere Wahrnehmungen, Analysen und Wertsetzungen zu verallgemeinern, und dem Bedürfnis, deren partikularistische Wurzeln gegen das Eindringen partikularistischer Wahrnehmungen, Analysen und Wertsetzungen anderer zu verteidigen, welche ihrerseits behaupten, Universalien vorzuschlagen. Wir müssen unsere Partikularismen verallgemeinern und gleichzeitig unsere Universalien partikularisieren, in einem ständigen, dia-

lektischen Austausch, der es uns erlaubt, neue Synthesen zu finden, die dann natürlich unmittelbar in Frage gestellt werden. Das ist keine einfache Angelegenheit (Wallerstein 2010: 60).

Enrique Dussel spricht auf ähnliche Weise vom Projekt der Transmoderne:

Wenn ich von Trans-Moderne spreche, beziehe ich mich auf ein globales Projekt, das die europäische oder nordamerikanische Moderne zu übersteigen sucht. Dies ist ein Projekt, das nicht post-modern ist, denn die Post-Moderne ist immer noch eine unvollständige Kritik der Moderne durch Europäer und Nordamerikaner. Hingegen ist Trans-Modernität eine Aufgabe, die in meinem Fall eben philosophisch zum Ausdruck kommt. Ihr Ausgangspunkt ist dasjenige, was verworfen, entwertet und als nutzlos bei den Kulturen dieser Erde beurteilt wurde, einschließlich der kolonisierten oder peripheren Philosophien. Dieses Projekt bezieht die Entwicklung des Potenzials jener ignorierten Kulturen und Philosophie mit ein, und zwar auf der Grundlage von deren eigenen Ressourcen, in konstruktivem Dialog mit der europäischen und nordamerikanischen Moderne (Dussel 2010: 63 f.).

Sowohl der »universale Universalismus« (Wallerstein) als auch das »Projekt der Transmoderne« (Dussel) kritisieren einen neutralen, als gegeben gesetzten Universalismus und weisen darauf hin, dass es sich auch bei universalen Ordnungsvorstellungen um fluide Gebilde handelt, die Ausdruck eines Prozesses sind, der Elemente der Einigung wie Elemente der Divergenz umfasst. Wallerstein und Dussel verfolgen dabei insbesondere das Ziel der Berücksichtigung der bislang durch den Kolonialismus und die Globalisierung Marginalisierten. Auch ohne ihren teilweise aufgeregten politischen Impetus übernehmen zu müssen, ist leicht einsehbar, dass eine tendenziell universalisierbare Ordnungsvorstellung mit Exteriorität wird umgehen müssen, ohne sie zu nivellieren. Trotzdem scheint ein gewisses Maß an Geschlossenheit geboten, um überhaupt von einem »Austausch« (Wallerstein) oder sogar von einem »Dialog« (Dussel) sprechen zu können.

Der konstruktive terminologische Vorschlag des vorliegenden Beitrages besteht darin, den Begriff der Integrität zur Kennzeichnung dieses Projekts in Anschlag zu bringen. Der Begriff der Integrität klingt dabei wie ein alter Vertrauter, weist aber bei näherem Hinsehen eine sträfliche Vernachlässigung innerhalb der Philosophiegeschichte auf. Dies ist einer der Gründe, warum er mir geeignet scheint, als (neues) normatives Paradigma zu fungieren – als eine

Begrifflichkeit, die verhältnismäßig wenig Ballast der Geschichte mit sich herumträgt.

Der zweite Grund ist ein höchst aktueller: Vor allem innerhalb des *environmental law* und der Debatten des Klimaschutzes lässt sich eine Zunahme der Verwendungshäufigkeit von *Integrität* beobachten. *Integrität* – so meine These – wird offensichtlich als angemessene Begrifflichkeit ins Auge gefasst, um drängende Fragen der Gegenwart konzeptionell zu erfassen. Ich möchte als *planetarische Integrität* Vorstellungen der Ganzheit der Erde verstanden wissen, die es als heuristische Konzepte deskriptiv allererst ermöglichen, einen Zusammenhang zu denken und als normative Paradigmen – sei es implizit oder explizit – eine Verpflichtung formulieren, die Erde als geteilten Lebensraum zu erhalten.

Planetarische Integrität zu Denken – und hier kommt die interkulturelle Philosophie ins Spiel – ist auf vielfache Weise möglich. Im Zusammenhang unserer Argumentation sticht freilich die Vorstellung von *Pachamama*, der Mutter-Erde, ins Auge, die über die Verfassung Ecuadors prominent Eingang in Recht und Gesetz gefunden hat. Der damit begründete *environmental constitutionalism* ist meines Erachtens zu Recht als Revolution gefeiert worden. Das Angebot lautet: Die Natur weltweit nicht mehr nur als Ressource zur Befriedigung menschlichen Interesses zu betrachten, sondern sie um ihrer selbst willen als Träger von Rechten zu bezeichnen. Auf diese Weise entsteht ein alternativer Denkraum für Recht und Moral, der uns das Gefüge von Ansprüchen und Verpflichtungen neu ordnen lässt. Dies stellt einerseits eine Art *Rehabilitation* der nicht-europäischen Kulturen dar, die durch das Projekt der Moderne marginalisiert worden sind, andererseits entsteht dadurch womöglich die normative Tiefengrammatik, um Probleme wie den menschengemachten Klimawandel und die Herausforderung aufzubauender Postwachstumsgesellschaften unkonventionell adressieren zu können.

Der menschengemachte Klimawandel ist keineswegs das einzige Problem, das die entstehende polyzentrische Weltgesellschaft herausfordert, es ist aber geeignet, um die Konvergenz zwischen westlicher Wissenschaft und nicht-westlicher Weltanschauung im Sinne der Integrität zu belegen. Zur Jahrtausendwende haben Geowissenschaftler um Paul Crutzen das Konzept des *Anthropozäns*[5] geprägt. Das Erdzeitalter, das den Namen des Menschen trägt, ist eine von weiten Teilen der Weltbevölkerung akzeptierte Redeweise geworden. Das Anthropozän liefert eine übergeordnete Beschreibungsperspektive, innerhalb derer das oftmals als Gegensatz beschriebene Verhältnis von Mensch und Natur, man möchte fast hegelianisch sagen, »aufgehoben« ist.

Das *Paris Agreement* wurde am 12. Dezember 2015 von 195 Mitgliedsstaaten der Vereinten Nationen unterzeichnet und sieht die Begrenzung der menschengemachten globalen Erwärmung deutlich unter 2 Grad Celsius vor. Abgesehen von der beanspruchten Sonderrolle der Vereinigten Staaten von Amerika, die Mitte 2017 ihren Ausstieg zum Jahre 2020 ankündigten, stellt die Übereinkunft von Paris der Form nach eine freiwillige Selbstverpflichtung der Staaten dar. Natürlich wissen wir, dass Papier geduldig ist und dass sich wohl weder abschmelzende Gletscher noch Winde und ozeanische Ströme an menschliche Vereinbarungen halten werden. Was mir als Philosoph zu untersuchen bleibt, ist die argumentative Struktur, auf die man sich dabei geeinigt hat. Die folgende Überlegung soll sowohl die darin enthaltene Reminiszenz an die ecuadorianische Verfassung zeigen und andererseits den Aufstieg der Integrität als Paradigma belegen: *»Noting the importance of ensuring the integrity of all ecosystems, including oceans, and the protection of biodiversity, recognized by some cultures as Mother Earth, and noting the importance for some of the concept of ›climate justice‹, when taking action to address climate change«* (Paris Agreement 2015, Präambel, Paragraph 13).

Die Präambel identifiziert eindrucksvoll die Integrität der Ökosysteme als neues normatives Paradigma im Konsens der Staaten der Erde und parallelisiert sie auf einen Vorschlag der lateinamerikanischen Staaten hin mit dem Gedanken der Mutter-Erde *Pachamama*. Drei weitere Male taucht der Begriff *environmental integrity* im Paris Agreement auf. Interessanterweise taucht *human dignity* kein einziges Mal auf. Die Erklärung von Stockholm (1972) kennt den Begriff Integrität (noch) nicht, ab der *World Charter for Nature* (1982) mit drei Nennungen und der *Rio Declaration* (1992) mit ganz und gar vier Nennungen scheint der Begriff Integrität salonfähig geworden zu sein.

Noch wissen wir nicht, ob Integrität tatsächlich der Beschreibungs- und letztlich auch Bewertungsmaßstab für die in Zukunft relevanten Fragen der globalen Gerechtigkeit sein wird. Im günstigsten Fall können wir gewissermaßen »live« mitverfolgen, wie sich ein normatives Paradigma herausbildet, das in einigen Jahrzehnten womöglich die Strahlkraft und Allgegenwart erlangt haben wird, die heute z. B. dem Würdebegriff zukommt.

Um eine Parallele zu ziehen: *Würde* ist bekanntlich seit Mitte des zwanzigsten Jahrhunderts in der *Allgemeinen Erklärung der Menschenrechte* (1948), dem *Deutschen Grundgesetz* (1949) und der *Genfer Konvention* (1949) prominent, wurde aber erst mit einer Latenz von rund 50 Jahren zum Gegenstand akademischer Forschungen. Diese Karriere, die der Würdebegriff im ausgehenden 20. Jahrhundert als Kristallisationspunkt historischer Erfahrungen und normativer Überzeugungen gehabt hat, regt uns an, *Integrität* ergänzend dazu zu überprüfen und die verheißungsvollen Versprechen, man bekäme damit etwa die globale Erwärmung, die große Transformation oder das Erdzeitalter des Anthropozäns besser in den Griff, kritisch zu hinterfragen.

4 Fazit

Mein konzeptionelles Angebot besteht darin, *Pachamama* und *Antropozän* als Spielarten *planetarischer Integrität* zu fassen, die innerhalb verschiedener Kulturen in der Lage sind, die Schutzwürdigkeit des Planeten weltanschaulich zu rechtfertigen und in einigen Fällen sogar juridische Rechte der Natur zu begründen. Die Verfassung von Ecuador (2008) kann als weltweites Gründungsdokument des *environmental constitutionalism* gelten. Das *Paris Agreement* führt die Integrität der Ökosysteme als parallele Formulierung der Mutter Erde ein und der Umstand, dass man 2017 auch in Indien zwei Gletschern und in Neuseeland einem Fluss Rechte zugeschrieben hat, zeigt vor allem eines: *Dekoloniales Denken* und *interkulturelle Philosophie* können hier erste Teilerfolge feiern, da hier nicht mehr Wissenschaft gegen Glaube, West gegen den Rest oder die Moderne gegen die Anderen ausgespielt wird. Innerhalb der Begründungen dieser Rechte tauchen neben wissenschaftlichen Erklärungen und Versatzstücken des internationalen Umweltrechtes vor allem auch indigene Vorstellungen der Maori in Neuseeland, der Hindu-Traditionen in Indien und der andinen Völker Lateinamerikas auf (Knauß 2018).

Planetarische Integrität ist meiner Ansicht nach intrinsisch verbunden mit dekolonialen Optionen und dem Projekt der Transmoderne. *Integrität* als Ergebnis eines durch die Exteriorität geläuterten Universalismus zu begreifen, manifestiert sich – so meine Hauptthese – nicht nur darin, indigene Weltanschauungen gleichberechtigt neben modernen Theorien der Wissenschaft stehen zu lassen und sie als begründende Narrative innerhalb einer globalen Rechtskultur zu akzeptieren, sondern auch darin, sowohl die Ausgeschlossenen als auch die ausgeschlossene Natur in den Kreis der Rechtsträger zu erheben. Nicht nur mit den *Verdammten dieser Erde*[6] – um eine Formulierung von Frantz Fanon aufzugreifen – gelingt es, die »verdammte Erde« normativ aufzuwerten. Die Rechte der Natur bilden das juridische Gegenstück zur planetarischen Integrität und weisen die Erde im Interesse der Menschen, aber nicht ausschließlich aufgrund menschlicher Interessen als schützenwürdig aus.

—Stefan Knauß, Universität Erfurt, Deutschland

Literatur

Bosselmann, Klaus, *The Principle of Sustainability: Transforming Law and Governance*, London etc., 2016.

Crutzen, Paul J., »Geology of Mankind«, *Nature*, Vol. 415 (6867), 2002, S. 23.

Crutzen, Paul J., and Stoermer, Eugene F., »The ›Anthropocene‹«, *Global Change Newsletter*, Vol. 41, 2000, S. 17–18.

Davison, Isaac, »Whanganui River Given Legal Status of a Person under Unique Treaty of Waitangi Settlement«, *The New Zealand Herald*, 15 March 2017. http://www.nzher ald.co.nz/nz/news/article.cfm?c_id=1&objectid=11818858 [26. April 2017].

Dussel, Enrique, »Eine neue Epoche in der Geschichte der Philosophie: Der Weltdialog zwischen philosophischen Traditionen«, *Polylog, Zeitschrift für interkulturelles Philosophieren*, Vol. 24, Wien, 2010, S. 47–64.

Fanon, Frantz, *Die Verdammten dieser Erde*, Frankfurt, 1966.

Husserl, Edmund, *Cartesianische Meditationen und Pariser Vorträge*. Husserliana I, 2. Auflage, Den Haag, 1963.

Knauß, Stefan, »Conceptualizing Human Stewardship in the Anthropocene: The Rights of Nature in Ecuador, New Zealand and India«, *Journal of Agricultural and Environmental Ethics* Vol. 31 (6), 2018, S. 703–722.

Lévinas, Emmanuel, *Totalität und Unendlichkeit – Versuch über Exteriorität*, Freiburg etc., 2002.

Martinez-Alier, Joan, »Environmentalism of the Poor«, in: *Handbuch Umweltethik*, ed. Ott, Konrad, Dierks Jan, und Voget-Kleschin, Lieske, Stuttgart, 2016, S. 224–229.

Merleau-Ponty, Maurice, *Phänomenologie der Wahrnehmung*, Berlin, 1966.

Merleau-Ponty, Maurice, *Das Sichtbare und das Unsichtbare*, München, 2004.

Quijano, Anibal, ›Coloniality of power and eurocentrism in Latin America‹, *International Sociology* 15(2), 2000, S. 215–232.

Weidtmann, Niels, *Interkulturelle Philosophie: Aufgaben – Dimensionen – Wege*, Tübingen, 2006.
Waldenfels, Bernhard, *Der Stachel des Fremden*, Frankfurt am Main, 1990.
Waldenfels, Bernhard, *Topographie des Fremden*, Frankfurt am Main, 1997.
Waldenfels, Bernhard, *Grundmotive der Phänomenologie des Fremden*, Frankfurt am Main, 2006.
Wallerstein, Immanuel, *Die Barbarei der anderen. Europäischer Universalismus*, Berlin, 2010.

Anmerkungen

1 *Pachamama* ist eine Gottheit, die von indigenen Völkern in den Anden verehrt wird. Der Begriff wird im Deutschen als »Natur« oder »Mutter Erde« übersetzt.

2 Als »Kolonialität der Macht« bezeichnet der peruanische Soziologe Aníbal Quijano (2000) die Ausprägung kolonialer Hierarchien auf der sozioökonomischen, politischen und epistemischen Ebene. Dekoloniales Denken beschreibt und kritisiert diese Machtbeziehungen.

3 In *Totalität und Unendlichkeit* (2002) führt Lévinas aus: »Dieser Begriff [Totalität SK] beherrscht die abendländische Philosophie. In der Totalität reduzieren sich die Individuen darauf, Träger von Kräften zu sein, die Individuen ohne ihr Wissen steuern. Ihren Sinn, der außerhalb dieser Totalität unsichtbar ist, erhalten die Individuen von dieser Totalität« (Lévinas 2002: 20).

4 »Die Weise des Anderen, sich darzustellen, indem er *die Idee des Anderen in mir* überschreitet, nennen wir nun Antlitz« (Lévinas 2002: 63).

5 Das *Anthropozän* ist eine neue geologische Epoche, innerhalb derer der Mensch als wichtigster Faktor für Veränderungen der Erde betrachtet wird. Die Geologen Paul J. Crutzen und Eugene F. Stoermer führten den Begriff im Jahr 2000 ein. Er löst das *Holozän* ab, eine post-glaziale geologische Epoche, die als Beschreibung für die vergangenen zehn- bis zwölftausend Jahre angenommen wird. Crutzen und Stoermer argumentieren, dass die »Aktivitäten der Menschheit sich allmählich zu einer bedeutenden geologischen und morphologischen Kraft entwickelt haben« (Crutzen und Stoermer 2000: 17). Die quantitative, wie qualitative Expansion der Menschheit in Bezug auf den Anstieg des Konsums und des Bevölkerungswachstums hat das Erdsystem deutlich geprägt. Insbesondere die Kohlenstoffdioxidemission bzw. deren Einfluss auf das Klima hat Crutzen zufolge das »Verhalten der Natur« verändert (Crutzen 2002: 23).

6 Der in Martinique geborene Franzose Frantz Fanon war Philosoph und Psychiater. In seinem Hauptwerk *Die Verdammten der Erde* (Fanon 1961) beschreibt er als politischer Aktivist unmittelbar vor seinem eigenen Tod die Dekolonialisierung Afrikas. Fanon hat in Algerien die Bemühungen der Lossagung von Frankreich begleitet und interpretiert den Prozess der Befreiung der Afrikaner als Vorgang der Subjektwerdung. Er ruft dazu auf, mittels eines gewaltsamen Kampfes gegen die Kolonialherren von »Dingen« zu Menschen zu werden. Fanons umstrittenes Werk teilt zudem den Pessimismus der Dependenztheoretiker, wonach die Menschen in den Kolonien »verdammt« bleiben, ein Leben in Abhängigkeit von den Wohlstandszonen zu führen, auch dann, wenn die unmittelbare koloniale Unterdrückung zu Ende sein sollte.

Socio-political Explorations

Environmental Ethics and Environmental Security: Specifics of the East-European Region

From the beginning of the 1970s, environmental ethics has claimed its status as a self-sufficient discipline developing a unique system of moral principles and imperatives for human behavior in the natural world.

The social significance of its scope is confirmed, in particular, by it becoming a part of study curricula. Environmental ethics has been taught in American universities since 1971—followed by educational institutions in Norway, England, Canada, and Australia. In the last quarter of the twentieth century, Club of Rome President Aurelio Peccei first introduced the concept of ecological culture. Since 1979, the journal *Environmental Ethics* has been published in the United States; since 1992, *Environmental Values* in Great Britain, and *Ökologie und Ethik* in Germany—currently followed by the journals *Ecophilosophy, Deep Ecologist, Between the Species*, and *Ethics and Animals*.

At that time, in the Soviet Union, environmental ethics as a discipline did not exist—and consequently, was not presented in the literature on ethics or in the tertiary curricula. Even now, introducing such new courses as Social Ecology, Environmental Ethics, or Bioethics is still considered as a future option in many universities for a number of reasons. For example:

As traditional ethics was defined as the normative regulation of human relationships with each other and social systems, it was considered that only people and society can be an object of moral attitude, with deforestation or cruel slaughter of animals excluded from the sphere of moral or immoral relations.

Countries with socialist principles of economic management were assumed to be reliably protected from global environmental crises.

Ecology was viewed through the lens of the class approach—and therefore, the recommendations of western environmental futurologists on anti-crisis preventive measures were perceived as ›enemy propaganda‹.

Marxist ethics did not accept that, to save the planet, we need fundamental changes in nature management, especially in socialist countries.

A dominant pragmatic approach to nature prompted society to engage in intensive economic research and transformation aimed at deriving economic benefits regardless of their consequences for nature. So the only ›environmental‹ slogan that was forming the environmental awareness of Soviet children, including future leaders of the Soviet society (from collective farm chairmen to General Secretaries of the Communist Party, who would make national decisions), was the quote of the last-century natural selectionist I. V. Michurin, »We must not wait for favours from Nature; our task is to wrestle them from her«. And today, this economic (not environmental) policy is implemented in the post-Soviet space as well—for example, extraction and sale of oil and gas in Russia, or the deforestation of unique forests in Belarus.

Today, at least in Belarus, environmental ethics faces some pressing challenges in implementing its regulatory function.

Environmental safety issues associated with the construction of a nuclear power plant in Belarus are problems not only for Belarus, but also for the European environment. Although the new plant has not yet been completed and commissioned, Nuclear and Radiation Safety has already been added to the International Sakharov Environmental University curriculum for training specialists for a nuclear power plant, including special courses on Risk Ethics and Management.

Moral-and-psychological problems related to the unresolved issues of rehabilitating people affected by the Chernobyl disaster. After more than thirty years since the catastrophe, many problems have been solved, and state and trans-national projects aimed at overcoming its consequences have been developed. There has been a large number of health studies of those affected by the accident, the extent of the economic damage has been calculated, and the ecological state of the region is being constantly monitored. However, the moral and psychological rehabilitation of the people and violation of associated eco-and bioethical principles are being lent much less importance or ignored. The problems of providing timely and reliable information, and the violation of the bioethical principle of ›informed consent‹ are primarily related to the moral-and-environmental responsibility of key decision makers.

Alternative renewable energy sources, a practical problem closely related to the previous two, is a focus of special attention of the whole world. It is in this direction that since 2004, the International Sakharov Environmental University has taught students majoring in Management of Renewable Energy Resources (in Belarus, this is first of all solar energy, wind energy, and energy derived from the biomass), and since 2008, in Energy-Efficient Technologies and Energy Management—as a result of the requirement to ensure the national energy security based on its own resources, and create the conditions for the maximum efficiency of using renewable energy resources while preserving nature, its balance, and biodiversity. These practical tasks have a significant ethical content *per se:* resolving them presupposes a certain level of environmental awareness of a specialist focused on the ability and willingness to observe such principles of ethics of the earth as ›chronological objectivity‹ and ›responsibility for the future‹ (Potter 1971: 205; Leopold 2013; Leopold 1992: 54–72), to use A. Leopold's terms.

Theoretical tenets of environmental ethics are applied by *providing biosecurity* related to developing science and modern biotechnologies. These have a significant potential of influencing people, society, and nature. However, these prospects are ambivalent. Given scientific and economic prospects of genetic engineering, we should also keep in mind potential risks of its further development. Today, the need for ethical regulation of scientific research becomes a practical task of environmental ethics, requiring an answer to the question whether we should always do what can be done in the field of biotechnologies.

Most ethical problems and disputes are associated with genetic engineering, human genetics, and GMOs. The latter are now actively used in pharmacology, agriculture, medicine (a number of vital medicines and vaccines), and food industry. Currently, transgenic (containing genetic material into which a DNA from an unrelated organism has been artificially introduced) plants can withstand the effects of many viral infections. At the same time, new biotechnologies, especially genetic-engineering technologies and nanotechnologies, may have potentially adverse effects on human health and environment.

As long as there is an element of scientific uncertainty about their possible adverse effects on human health and the environment, according to the ecological precautionary principle, these new forms of technological activity should be regulated at the state level by special legislation. In Belarus, this is the 2005 Law on Biosafety, which enables a thorough analysis of GMOs. At the same time, we need to conduct serious ethical-and-educational work aimed at breaking stereotypes and prejudices towards GMOs and GMPs.

For *safe medicine,* we need ethical regulation and governance of fundamentally new areas of transplantology, reproductive technologies (artificial insemination, IVF, embryo transplantation), birth control (abortion and contraception), genomics (genodiagnostics, genetic engineering, gene therapy, genetic personal identification), psychopharmacology, and reanimatology. On one hand, the expansion of experimental research in the areas with direct access to medical practice has created new opportunities of genetic manipulation, fetal surgery, new technologies of childbirth, and organ transplantation, maintaining the patient in an unconscious vegetative state for a long time. On the other hand, this research and practical success have created non-standard situations aggravating ethical and legal problems of determining the legal status of the embryo and the legal bases for distributing donor resources during transplantation, developing new approaches to defining the criteria for death, controlling the development of genetic engineering and biotechnologies to prevent their catastrophic consequences for human species. Such forms of genetic intervention in the human body as genetic therapy of embryo and somatic cells, and obtaining identical genetic copies of an organism require ethical evaluation and discussion of their consequences, as the associated decisions will affect the research direction and rate. We also require ethical assessment for forming an adequate societal response to the use of the research results.

Thus, the practical outcomes of environmental ethics, especially in the field of safe medicine, are directly related to the new theoretical direction of *human ecology* on the cusp of ecology and environmental hygiene (ecological medicine), which coordinates and balances public health with ecosystem functioning. Human ecology studies objective laws of functioning, development, and interrelation of the integrated dynamic system ›environment—human‹, where the priority of protecting human health should take into account its bio-psycho-social essence, moral values, and guidelines—in coordination with the principles of organisation and functioning of biological systems of the natural environment.

Human ecology aims at ensuring the environmental safety of the population, but the most important factor here is the ethical aspects of our behavior in nature—our moral attitudes, values, priorities, and the level of moral culture. Currently, as our economic activity poses ecological danger, it would seem logical to assume that to ensure environmental safety, we would need to ban, minimise, or limit our economic activity. But since this is impossible, a prerequisite for regulating the environmental impact of natural objects on humans is our compliance with the *moral imperative* of the conscious limitation of our negative impact on the environment. According to the founder of bioethics R. V. Potter, the introduction of moral regulations by environmental ethics and bioethics can help resolve this ›conflict of interest‹, becoming a ›bridge to the future‹ for the humanity to survive.

Unlike the rest of the biological world, we have found ourselves outside the factors of natural selection, having entered the social evolution without being sufficiently adapted genetically to the adverse effects of the environment. Moreover, progressive technological advances of the environment negate the natural selection of new genetic adaptations, and weaken genetic reserves inherited from previous generations, while fixing negative and pathological mutations in the human genome—with the constantly increasing anthropogenic flow of chemical compounds (artificially synthesised substances alien to the biosphere). In small doses, the mutagenic effect of these factors seems not to change human health, and goes unnoticed, but the cumulative negative effect of these factors for subsequent generations will be reproduced, like any other inherited characteristics, indefinitely. There is no straightforward solution of these issues in eco- and bioethics. Based on the ethical principles of pragmatism and utilitarianism, adherents of the ›liberal‹ approach (molecular biologists, geneticists) see significant perspectives in genetic therapy and biotechnology, viewing any limitations as an obstacle to scientific progress. Humanitarians mostly hold the ›conservative‹ point of view, being concerned about possible genetic changes which, once started, can unrecognizably change the genetic portrait of humanity. The ethical arguments of this position are related to the expectations of catastrophic consequences as a result of scientific researchers refusing to observe traditional moral norms. From the point of view of researchers in biology, the liberal approach is a better fit for the modern state of science and society, and their progress.

This position seems to be fully justified. Adverse environmental conditions and a number of other causes lead to the increased number of children born with serious hereditary defects. For most of them, there is no effective treatment yet—but there is a real diagnostic potential of detecting many genetic diseases at the embryo or fetal stage. It will soon become possible to correct the genetic code, and optimise the genotype of the unborn child—enabling us not only to avoid many genetic conditions, but also improve physical, psychological, and mental attributes of newborns. So far, the most common reaction to diagnosed genetic diseases at the embryo or embryonic stage is abortion, the destruction of a potential life.

Thus, the ethical problems of human ecology are most directly related to human rights—first of all, *the right to life* and as a result, the still disputed *right to death*. It is the ethical position of specialists and general public that currently determines the societal attitude towards euthanasia and organ and tissue transplantation.

Another crucial ethical problem related to conserving natural biodiversity for its sustainable development is the *ecology of animals and the rights of nature*. Modern culture has shown the danger of treating humans as a self-purpose. It is not humans within the framework of traditional humanism and anthropocentrism, but their choice of genuine values that opens a new post-humanity value—a higher level of humanity, revealing the ability not only to be self-centred, but also to care about the life and rights of anything living, taking humanity beyond the human as a biological species (Tulchinsky 2002: 211).

1 Environmental Ethics in the East-European Region

As part of creating conditions for sustainable development in the East-European region, environmental ethics now includes objects of nature as its subjects. This has led to environmental ethics which develops new moral standards of individual behaviour in ›human‹ and ›inhuman‹ situations, and through its universal character, permeates every aspect of our lives.

Therefore, the question whether environmental ethics is a set of theoretical principles or an empirical result and guideline for life practice is, in my opinion, just a play on words and semantics. It is both. Environmental ethics builds its own problem scope and system of moral categories and basic principles on theoretical and applied levels. At the theoretical level, we develop conceptual foundations of environmental morality as a mandatory sphere; also, we search and justify fundamental values, their criteria, and posit the theoretical principles of eco-ethical regulation. At the applied level, we formulate specific rules of moral judgement to answer the practical question ›What should I do?‹, and regulate human behaviour in its natural habitat. Both levels are intertwined and inter-related in solving the following main tasks of environmental ethics:

- identification and theoretical substantiation of the basic principles and norms of moral regulation of the relationship between man and nature;
- search for practical regulations, technologies, and institutions of their interaction;
- development of tools and methods for forming environmental consciousness, which synthesizes a global vision of the world with truly humanistic values, and is the basis for the ethical human behaviour in practical activities.

So, specific conditions for sustainable development in each country determine priorities in addressing these challenges.

The studies of the scientific foundations of environmental ethics in the East-European region of the post-Soviet space started at the beginning of the twenty-first century with academic discussions on differences and specifics of applied, professional, and practical ethics (Apresyan and Guseynov 2002: 389; Guseynov and Kon 1989: 278–279; Bakshtanovsky and Sogomonov 1999: 154; Guseynov 2004, 148–159; Apresyan 2004: 160–181). It is within the framework of *applied ethics* that Professor R. G. Apresyan (Institute of Philosophy, Russian Academy of Sciences (IP RAS) proposes to analyze environmental ethics as a set of systematic and justified moral ideas about the relationship between human and nature. In 2004–2007, within the framework of the projects HESP and *Improving Teaching of Ethics in the Higher Education System*, the Department of Ethics at the IP RAS organized seminars and summer international workshops for young Russian teachers of ethics in Kaunas, Kiev, Minsk, and Moscow region—taught by well-known specialists in ethics Holmes Rolston III, T. Gouvier, A. A. Huseynov, B. G. Yudin, P. D. Tishenko, and experts from other republics of the former Soviet Union. This work resulted in publishing a collection of scientific reports *Ethics and Ecology*, with the support of the UNESCO office in Moscow (Apresyan 2010).

Currently, a number of centres for environmental ethics are being set up in Russia. Scientific-and-Research Institute for Applied Ethics at the Tyumen State Oil-and-Gas University has been for many years most influential in studying *the theoretical issues* of applied ethics, including environmental ethics. *Vedomosti*, its regular publication edited by V. I. Bakshtanovsky and Y. V. Sogomonov, reflects the development of ethical thought not only in Russia, but in the entire post-Soviet space (Bakshtanovsky and Karnaukhov 2009: 252).

Under the sponsorship of IP RAS and the UNC Center for Bioethics, Ogarev Mordovia State University (Saransk) is becoming a centre for developing a methodology for teaching environmental ethics (with the contribution of A. A. Sychev and M. D. Martynova).

Philosophical and methodological problems of forming an ecological world outlook are the focus of attention at the Mytischi Branch of Bauman Moscow State Technical University. Environmental ethics is discussed at the annual International Scientific Conferences Eco-World (under the guidance of V. I. Falco) and in the journal *Forest News*.

In Russia, the issues of environmental ethics are analyzed not only by the state, but also by NGOs—for example, the international social group ECOERA (Skolkovo) holds regular forums under the title *Ecology for Life: Integration and Development for Future Generations*, with the presentation of the ERAECO National Environmental Award.

Various aspects of environmental ethics are being studied also in other countries of the former Soviet Union. In the Ukraine, the Kiev Ecological and Cultural Centre was one of the first in the post-Soviet space to address the issues of environmental ethics and animal rights. Since 2000, under the leadership of its director V. E. Boreiko, the honoured nature conservationist of Ukraine, the Kiev Centre has held a series of international seminars on environmental ethics. For more than 15 years, it has been publishing the international *Humanitarian Environmental Journal*, a mouthpiece of environmental ethics for the post-Soviet space. The journal published Boreiko's popular brochures in the Wildlife Conservation series (Boreiko 2004a, 2004b) and five editions of his book *Breakthrough in Environmental Ethics* (Boreiko

2013). At Professor Boreiko's initiative, Ukraine adopted the *Law on the Protection of Animals from Cruelty* in 2006 (the only one in the post-Soviet space) which has put environmental activities in Ukraine on a legitimate basis, in particular, making it possible to bring (and win) lawsuits for crimes against the rights of nature.

In the Republic of Moldova, environmental ethics issues are researched at the Nicolae-Testemiţanu State University of Medicine and Pharmacy within the framework of the project *Survival Strategy in the Context of Bioethics, Anthropology, Philosophy, and Medicine* led by T. N. Tsyrdya, the developer of the concept of social bioethics. T. N. Tsyrdya has initiated and organised annual international scientific conferences in Chişinău, and published its proceedings in Romanian, English, and Russian (the twenty-fourth compendium was published in 2018; see Tsyrdya 2018).

We should also note the significant contribution of Lithuanian ethics experts to developing environmental ethics in the post-Soviet space—with the associated terminology of organisational ethics, professional ethics, company ethics, and institutional ethics. In particular, N. Vasilievene introduced the concept of institutional ethics, which develops the principles of ethical activities not for people, but for organisations. The institutionalisation of eco-ethics, its transformation into an effective tool of economic management, and the development of environmental management mechanisms are some of the most vital and poorly-studied areas. Lithuania (in particular, N. Vasilievene and V. Motikeitite from the Mykolas Romeris University, Faculty of Strategic Management and Policy; see Vasilievene 2010) is the undisputable leader in this space. Institutional ethics is now promoted by Ethical Committees and Ethics Commissions—primarily at medical institutions, but also in parliaments, journalists' groups, religious organisations, and company Boards—and their principles are reflected in the Codes of Ethics of these entities. These activities of developing environmental ethics in, and for, the post-Soviet space have resulted in the publication in Lithuania of the collective monograph *Environmental Ethics: the Power of Ethics for Sustainable Development* (Vasilievene and Jurciukonyte 2010).

The spontaneous ›division of labor‹ in environmental ethics has become, in our opinion, highly productive, with the issues of forming environmental consciousness (which, we think, can be effected only by continuous all-out ecological and ethical education) being studied at the Mordovian State University in Russia (M. D. Martynova) and the International Sakharov Environmental University in Belarus (T. V. Mishatkina). We consider environmental education to be the most important practical component and function of environmental ethics.

2 Principles, Norms, and Regulatives of Environmental Ethics

In the post-Soviet space, the norms and values of environmental ethics are primarily revealed through specific differences between anthropocentrism and non-anthropocentrism: it was through the lens of the human-nature relationship that these differences were identified and analyzed. *Anthropocentrism* is defined through the predetermined attitude of human beings towards nature, the concern for our own needs and interests, with nature perceived only as a necessary condition for our existence. In contrast, *non-anthropocentrism* recognizes the intrinsic value of nature as such, regardless of its benefit or ›harm‹ to humans. In different theories, this intrinsic value is attributed to biological species, or individual entities, or only to sentient (and suffering) animals, or ecosystems. Non-anthropocentrism as a moral value and ethical principle is a normative and ethical innovation which extends the subject of moral responsibility to non-human lives and communities and ecosystems—without diminishing the role and significance of humans, but rather postulating a harmonious and equal community of people and all other components of nature as the highest level on the value scale.

Values of nature is another debatable issue. They are determined by an anthropocentric or non-anthropocentric position, and are significant for identifying eco-ethical principles: should we recognize the independence and inherent value of natural objects or determine them depending on human needs and interests? This is not a theoretical or scholastic question, but rather a practical eco-ethical one. It is the foundation stone of the whole issue of equal rights of living entities to life and compassion. Some environmentalists believe that ›no single living being, except man, possesses the »inherent value« of itself‹, using the absence of interests, desires, and will of natural objects (which, by the way, is arguable) as an argument (Marfenin 2007). This argument would have been valid for *the awareness of* this inherent value (it is impossible for non-humans)—but we are talking about something else, about its objective presence or absence, regardless of us being aware of it. We believe that ecosystems should be recognized *as* independent moral subjects with their inherent inalienable value, regardless of their ability to be aware of their self-worth (the baby does not have this awareness either, but its life is still self-valuable)—we as humans must exercise this awareness for them. We do not have the right to make decisions on the species value or their right to life. For the sake of biodiversity, we must strive to preserve all species and natural objects. Thus, eco-ethics imposes a regulatory and moral moratorium on treating nature as a thing and a resource, and it is only the self-value of natural systems that can form the basis of modern eco-ethics.

A *moral-and-understanding attitude to nature*, to use V. A. Petritsky's term (see Petritsky 1990: 103–106; Potter 1971), is another attribute of the non-anthropocentric approach, manifesting itself in the aspiration to ›spiritualize‹ and understand the Living Being, even to the point of penetrating its feelings and experiences. To feel such empathy, it is necessary for non-human living subjects to be recognized as equal to human subjects, and this is possible only through forming subject-subject *relationships* between man and nature, with the world of natural phenomena perceived as its ›Other‹, Other entity, Other sentient being or social organism. This concept has been developed by Vladimir Falco (see Falco 1991: 23–29).

Professor Vladimir Falco, an Associate Member of the Russian Ecological Academy and International Informatisation Academy, has published more than 40 works on the philosophical aspects of ecology, globalisation, environmental ethics, and bioethics. He is an organiser of EcoWorld international conferences (2009–2019), and is a recipient of the Lomonosov Gold Medal *For the Outstanding Contribution to Science and Ecology.*

Subject-subject relationships of human beings and nature determine and enable the formation of moral values of eco-ethics based on the feelings of love and compassion for nature and the feeling of urgency, when we are concerned about the natural conditions for future generations (Leopold 1983, 1992). Our thoughts about the future, which we owe to our descendants, should be based on specific moral principles:

- *principle of chronological objectivity*, which prohibits ignoring the interests of people based on their temporal, spatial, or ideological distance;
- *moral imperatives of our dialogue* with the future, which imply

the rejection of any actions undermining the possibility of the existence or interests of future generations.

Following these norms is possible only under the conditions of moral and environmental freedom and responsibility. Their ratio is determined by the degree of knowledge of social and natural laws, and the possibility of mastering and ›manipulating‹ them. Obviously, this freedom depends on following the fundamental tenets of moral and environmental *responsibility,* including a transition from the ›model of predominance‹ over nature to the ›model of coexistence‹ with nature; an adoption of a new concept of environmental protection not so much *for* humans but rather *from* humans; and finally, a ›reconciliation‹ of economics and economy with ecology, based on moral criteria.

In this respect, the *Environmental Ethics* document developed by the UNESCO expert group is of particular interest. In 2004–2006, a group of experts developed a policy document on environmental ethics (discussed at the Fourth and Fifth Ordinary Sessions and intermediary extraordinary session—but never published) which made a substantial contribution to understanding environmental ethics as a set of principles. Specifically, some of the principles were: respect for all life, respect for biodiversity, safeguarding the sustainability of the biosphere, environmental justice, precautionary principle, Earth as global commons, rights of future generations, shared responsibility, contraction and convergence, principles about war and the environment.

UNESCO independent expert groups (including Russian representatives as well) research and develop regulatory documents on bioethics, ethics of science, and ethical issues related to the global climate change. Within the framework of the UNESCO programme, R. G. Apresyan has developed general and practical principles of environmental ethics (see Apresyan 2011). Prof Apresyan, Doctor of Philosophy, Head of the Ethics sector of the Institute of Philosophy of the Russian Academy of Sciences, International UNESCO Expert, proposes to view environmental ethics as a set of systematised and reasonable moral ideas about the relationship between humankind and nature.

His *general principles of environmental ethics* include respect for all forms of life, biodiversity, maintaining the sustainability of the biosphere, environmental justice, the precautionary principle, and common heritage of natural resources. The principle of *respect for all forms of life* affirms the value of Life in the spirit of Albert Schweitzer's *reverence for life,* the moral significance of every living being irrespective of human interests (see Schweitzer 1992: 7).The principle of *biodiversity* affirms its value and the need to preserve it as a manifestation of the richness of nature. It is not the same as the principle of respect for all forms of life. Any force should be used only as a last resort, in response to a clear, immediate, and imminent threat, and the use of force should be commensurate with the degree of threat. The principle of *maintaining the sustainability of the biosphere* could be even more important than the preservation of individual life, species, or ecosystem (except humans). With the global climate change calling into question the existence of humankind, this fundamental principle is underlying the concept of sustainable development, and its priority is being increasingly recognized by the UN and international communities. The principle of *environmental justice* affirms equal distribution of the right to environmental safety among people, with each of us being responsible for its preservation. We know that environmental degradation and natural disasters usually have more acute negative consequences for the most socially and economically vulnerable groups of the population. According to the *precautionary principle,* in developing policies with humanitarian and environmental consequences, we should first of all take into account the most dangerous worst-case scenario. When human activity can lead to morally unacceptable damage, when there is a risk to life or health, when there are potential irreversible consequences negatively affecting future generations or the environment, we need to target our efforts at its reduction or prevention. The principle of *common heritage of natural resources* is based on the idea of our planet as a whole entity. We are equally responsible for natural resources and the environment, as depleting natural resources will affect all people, now and in the future. The commonality of resources is an ethical, not a legal category. Although it is not possible to enact this idea in a law, the *Rio Declaration on Environment and Development* (1992) postulates the development of associated international legal mechanisms.

Apresyan's *practical principles of environmental ethics* include rights of future generations, shared responsibility, ›presumption of danger‹, reduction, and convergence. According to the principle of the *rights of future generations,* current population should take care of future generations of people and other living beings. Although we do not know potential needs of future generations, in any case, they will want to live in a favorable environment, with a stabilized world population, and sufficient reserves of food and fresh water. The principle of *shared responsibility* states that since natural resources are common property, the responsibility for environmental protection should be shared by all, and not be delegated to any particular organisation, group, or country. The principle of ›*presumption of danger*‹ (based on the respect for life and the precautionary principle) is expressed in the practical requirement for those who undertake actions with environmental consequences to bear the burden of proving their safety. The principle of *reduction and convergence* (based on the environmental justice and common heritage of natural resources principles) is related to the emission of gases into the atmosphere, the accumulation of which aggravates the greenhouse effect. According to this principle, we need to comply with quotas restricting the volumes of emitted gases—as per the Kyoto Protocol and the Intergovernmental Agreement on Climate Change.

The above principles, norms, and imperatives may well become *conceptual foundations of environmental ethics*, and contribute to the process of ecologising morality. They may include:

- determining our attitude to natural objects not only by material, economic, legal, or administrative prescriptions, but also by moral norms and principles;
- ecologising ›traditional‹ ethical norms and principles, forming new ›ethical-and-bio-ecologised‹ moral values (›environmental conscience‹), and minimising ›old‹ moral values of usefulness and expediency;
- forming a unified moral and environmental responsibility in economic, professional, and domestic nature uses;
- observing the ›ecological imperative‹ of N. Moiseyev, which demands/orders people to accept equal vulnerability of humans and natural environment, without exceeding their ›limits of strength‹ or conflicting with the laws of nature (Moiseyev 1991). Professor Moiseyev, Academic of the Russian Academy of Sciences, one of the thought leaders in ecophilosophy, has introduced the concept of ecological imperative—a set of rules and restrictions for humankind to be able to secure a future for itself.

3 Ecological and Ethical Education in Belarus

Ecological education, forming a new type of ecological consciousness and values system, plays a leading role in creating awareness of environmental problems, contributing to establishing a rational and careful attitude to the environment, and forming a socio-intellectual base for implementing co-evolution and sustainable development. The requirement for the societal ecological culture is reflected in the Constitution and the law *On Education* of the Republic of Belarus. They define the principle of environmental orientation as a key one. The law *On Environmental Protection* provides for obligatory environmental studies in all educational institutions and forms of education (Article 75), and stipulates that the officials involved in environmental management should have had the associated environmental training and regularly update their knowledge. These requirements are reflected in the *National Strategy for Sustainable Development Until 2020.*

The main factors determining a *conceptual model of environmental education in Belarus* are:

- relevance and necessity of solving environmental problems on global and regional scales as a universal factor. In Belarus, this factor is complemented by the regional factors of the post-Chernobyl situation;
- low level of ethical-and-ecological awareness and culture of the general public, professional ›nature users‹, and decision-makers —as a universal factor;
- in an environmental crisis, the awareness of the dependence of environmental solutions on the mentality and the level of the societal ecological culture;
- theoretical issues related to developing the concept of ecoethics discussed above.

Relatedly, some fundamental features of ecological education are:

a. indissoluble unity of rational and logical, analytical, and emotional and sensual origins—and hence empathy, appeal to the feelings of love, pity for natural objects needing our care and protection, ›don't be afraid to be ridiculed for sentimentality‹—in A. Schweitzer's words;

b. continuity of the environmental education, using temporal and spatial aspects of the ›always and everywhere‹ principle;
c. non-trivial forms and methods of training (case study).

Currently, ecological education in Belarus is aimed at ensuring its continuity: it should start in primary school. Integrated environmental courses are now part of the curricula at all educational establishments, with specialized environmental groups set up at some of them, and some environmental aspects being taught even at some preschools. Extracurricular institutions have accumulated considerable experience in environmental education; their work in this area has been managed by the Republican Environmental Centre for Children and Youth focusing on active practical work on environmental protection.

Secondary colleges have added special courses on the fundamentals of ecology with elements of eco-ethics to their curricula, implementing an inter-disciplinary approach in the environmental training of specialists, with some environmental issues reflected in graduation theses.

In tertiary education, environmental courses (Ecology, Radiation Ecology, Environmental Protection and Rational Use of Natural Resources, Radiobiology and Radiation Medicine, Environmental Economics) are a mandatory part of the curricula, with special scientific-and-methodological committees responsible for developing educational standards. Environmental aspects are an important component of the course Protection of the Population and Economy in Emergency—Anti-Radiation Safety, which is taught at all tertiary schools.

Environmental specialists in Belarus are trained at all state universities: the Belarusian State Technological University, Belarusian National Technical University, Belarusian Agrarian-and-Technical University, Belarusian State Agricultural Academy, Polotsk State University, and Brest Technical University.

The *International Sakharov Environmental University* (ISEU) set up in Minsk in 1992 in response to the Chernobyl challenges, plays a special role in forming and developing environmental education in the East-European region. It was A. D. Sakharov who shortly after Chernobyl suggested opening universities in the three countries most affected by the terrible catastrophe (Russia, Ukraine, and Belarus), which would focus on training specialists in preventing such catastrophes, dealing with their consequences, and solving radio-ecological, radiobiological, and technological problems. Such a university —named after A. D. Sakharov—was opened only in Belarus, after his death.

Andrei Dmitrievich Sakharov was a Russian nuclear physicist, dissident, Nobel laureate, and activist for disarmament, peace, and human rights. He became renowned as the designer of the Soviet Union's RDS-37, a codename for Soviet development of thermonuclear weapons. Sakharov later became an advocate of civil liberties and civil reforms in the Soviet Union, for which he faced state persecution; these efforts earned him the Nobel Peace Prize in 1975. The Sakharov Prize, which is awarded annually by the European Parliament for people and organisations dedicated to human rights and freedoms, is named in his honor.

Using the interdisciplinary focus as the most promising way of scientific research, ISEU was the first in Belarus to set forth and implement the idea of teaching environmental ethics in an environmental university.

ISEU is leading the Educational-and-Methodological Association of Tertiary Schools of Belarus in ecological education, and the Permanent Commission on Radio-Ecological Education of the CIS countries. It is constantly upgrading its environmental training, introducing new disciplines, such as Radio-Ecology (being now involved in IAEA activities), Nuclear and Radiation Safety (as part of the state training program for nuclear power plants specialists), Management of Renewable Energy Resources, Information Systems and Technologies in Ecology, Medical Ecology.

The multidimensional tasks of training environmental specialists have resulted in developing a special *ecological education programme* at ISEU—introducing a mandatory 20-hour course Fundamentals of Environmental Ethics (using traditional and non-standard interactive methods of training) aimed at forming conscious moral attitude towards nature, the Other Living, an active and reverential attitude towards Life in general, and readiness to be guided by the principles and norms of the environmental ethics.

To provide professional methodological support for the programme, we have created a network of experts to develop curricula for continuous ecological and bioethical education, publish textbooks, manuals, and guidelines in Environmental Ethics and Biomedical Ethics. All educational institutions have received a video-pack of training materials on ecological and bioethical education. One of the most important achievements of ISEU is the publication and free distribution of study materials on environmental ethics to Belarusian educational establishments (Mishatkina and Kundas 2008; Mishatkina 2008; Mishatkina and Melnov 2008; Mishatkina 2011; Kundas and Mishatkina 2011; Mishatkina and Melnov 2011; Mishatkina and Melnov 2018).

We are also involved in the public approbation and discussion of the issues of environmental and ethical education in Belarus. ISEU holds national and international conferences, seminars, and summer schools for teachers and students. Annual international scientific and practical conferences *Sakharov Readings: Environmental Problems of the Twenty-First Century,* with the participation of experts from Russia, Ukraine, Moldova and other CIS countries, publish a special issue (based on the conference materials) of the *Ecological Bulletin.*

With mastering the basics of environmental ethics by humanising the content of special study subjects, some parts of eco-ethics and bioethics are integrated into Radiobiology and Environmental Medicine subjects. Experiments and tests on animals have been replaced with alternative methods, such as the optional Alternatives to Using Test Animals in Education course based on computer training programs.

Environmental education has expanded its scope to forming the personality of a specialist in extracurricular time—through participation in conferences, round tables, and press conferences, including the commemorative *Sakharov Readings: Environmental Problems of the Twenty-First Century.*

In 2005, the students initiated the *EcoUni Association,* which is aimed at involving students in educational and outreach activities related to ecology and environmental protection. Its educational project *Man–Ecology–Bioethics* has contributed to developing the environmental curricula for schools and pre-schools. It was awarded a diploma, and won the National Environmental Award. Its programs were included in the UNESCO-Project *Environmental Ethics in the System of Bioethical Education of the Republic of Belarus.* Based on their experience, EcoUni members prepared a *Bioethics and Eco-Ethics for Schools and Extracurricular Education* study guide (later expanded to include a special *Alternatives to Animal Experiments* content for the students of biomedical disciplines). They also held an international seminar on protecting animals in education and research with the title *Humanisation of Medical and Biological Education.* Participants included the InterNICHE (United Kingdom), Centre for the Protection of Animal Rights Vita (Russia), and the National Committee on Bioethics of the Republic of Belarus. EcoUni is now InterNICHE's official partner in Belarus, having received its grant to implement a project on humanising education.

As part of extensive environmental education, ISEU students volunteer for ›ethical and environmental patronage‹ in schools, make films and videos, take part in environment-themed film festivals, and contribute to the national magazines *Ecologist and I* and *Animal World*. The meetings of their DISK discussion club focus, *inter alia*, on GMOs and ethical problems of modern biotechnologies. Participation in the annual almanac *Poetic Sakharisation* demonstrates the students' ability to see the world in a non-trivial light, which is the *sine qua non* for the modern environmental thinking.

ISEU environmental and educational activities have received international recognition—in particular, within the framework of its cooperation with UNESCO. Under its aegis, ISEU implemented the *Environmental Ethics in the Educational System of the Republic of Belarus* project, which has transformed its Volma training and conference complex into an *environmental information and education centre* for Belarusian and CIS student and post-graduate advanced training seminars and workshops. With UNESCO's assistance, the Centre has set an *ecological park*—with a demonstration platform holding installations showing the use of renewable energy sources, and an ›ecological path.‹ Also, the Centre has prepared a *Park Guide*, and trained students as guides. The students' immersion into the information and educational environment that combines clean natural environment, the latest educational and energy technologies, ethical and humanistic training, and national cultural and historical traditions contributes to forming an environmentally responsible self-developing person.

The Centre's inauguration included the Republican seminar and teacher-training workshop *Ecological Ethics and Human Ecology* (in subsequent years, followed *by training seminars on environmental ethics* for university staff and young people of Belarus). This opening presentation marked the beginning of the Centre becoming also a methodology hub for public discussions of topical social and environmental problems through Summer Schools of a Young Ecologist, and conferences and training seminars for target audiences.

The latest introduction in the curriculum, the development and implementation of *the ISEU Transformation Programme into a Climate-Neutral University*, is our response to the UN call for climate neutrality. As an environmental university, ISEU views this challenge as a ›guide to action‹, focusing on education and research in the field of renewable energy sources; environmental education, and administrative and organisational measures. We have completed all stages of the programme, using its results not only at the university, but also externally, and distributing the published materials to the educational establishments of Belarus and CIS. We share our experience in the ecological education at international conferences and seminars.

So, for the first time in the post-Soviet space, we have developed a *conceptual model of environmental ethics* based on the non-anthropocentric paradigm of the human-nature relationship and including ethical principles, norms, and standards complying with relevant international documents and policies. Also, we investigated the structure, functions, and basic concepts of eco-ethics, and identified bioethical problems of human ecology. We direct our sustainable development strategy of its expansion to the educational system at introducing and strengthening international standards in solving ethical problems resulting from the progress in science, technology, and biotechnologies, as well as the national and international environmental situation.

4 Summary

Considering the determining factors, conditions, and specific features of the formation of environmental ethics and environmental-and-ethical education in the East European region, I have analyzed key theoretical principles that need a fundamental justification in order to develop a conceptual model of environmental ethics and environmental-and-ethical education. The paper presents and defines basic principles, rules, and regulations of environmental ethics, as well as the most controversial »open« problems of environmental ethics, *i. e.* anthropocentric and non-anthropocentric, subject-objective and subject-subjective attitude towards nature, recognising the rights of nature etc. I suggest the conceptual model of environmental ethics, which makes it possible to address the issues of environmental-and-ethical education in the post-Soviet space, under the conditions of the authoritarian approach to solving environmental problems and low levels of environmental culture, in particular, in the Republic of Belarus.

—*Tatiana Mishatkina, International Sakharov Environmental University, Minsk, Republic of Belarus*

References

Apresyan, Reuben, »On Professional Ethics«, in: ed. *Professional Ethics*, V. 1, ed. Vladimir Bakshtanovsky and Nikolay Karnaukhov Tyumen, 2004, pp. 160–181.

ed. Apresyan, Reuben, Ethics and Ecology, Veliky Novgorod, 2010.

ed. Apresyan Reuben and Guseynov, Abdusalam, Professional Ethics/Ethics: Encyclopaedic Glossary, Moscow, 2002.

Apresyan, Reuben, »Main Principles of Environmental Ethics«, in: Environmental Ethics, ed. Tatiana Mishatkina and Semyon Kundas Minsk, 2011, pp. 5–20.

ed. Bakshtanovsky Vladimir and Karnaukhov, Nikolay, Paradigms of Applied Ethics, Vol. 35, Tyumen, 2009.

Bakshtanovsky Vladimir and Sogomonov, Yuri, »Ethical Applied Identification of Environmental Ethics: Invitation to Reflexive Self-Determination«, in: *Environmental Ethics. The Power of Ethics for Sustainable Development*, ed. Nijole Vasiljeviene and Agne Jurciukonyte Vilnius, 2010, pp. 83–110.

Boreiko, Vladimir, Abridged Course of Environmental Ethics, Kiev, 2004a.

Boreiko, Vladimir, Environmental Ethics in School, Kiev, 2004b.

Boreiko, Vladimir, Break-Through into Environmental Ethics, Kiev, 2013.

Falko, Vladimir, »Ethical Prospects of Ecological Consciousness«, in: Ecological Issues Under the Conditions of Perestroika, ed. Vladimir Falko, Moscow, 1991, pp. 23–29.

ed. Abdusalam Guseynov, and Kon, Igor, Professional Ethics/Ethics Glossary, Moscow, 1989.

Abdusalam Guseynov, »On Professional Ethics«, in: Professional Ethics, ed. Bakshtanovsky, Vladimir and Karnaukhov, Nikolay, Vol. 25, Tyumen, 2004, pp. 148–159.

ed. Kundas Semyon, and Mishatkina, Tatiana, Environmental Ethics, Minsk, 2011.

Leopold, Aldo, *A Sand County Almanac*, Moscow, 1980.

Marfenin, Nikolay, »Ecology and Ethics«, in: Russia in the Environment, ed. Marfenin, Nikola,y Moscow, 2007, pp. 166–185.

Moiseyev, Nikita, Man, Environment, Society, Moscow, 1991.

ed. Mishatkina, Tatiana, Bio-Ethics and Environmental Ethics for Secondary and Extracurricular Education, Minsk, 2008.

ed. Mishatkina Tatiana, and Kundas, Semyon, Fundamentals of Environmental Ethics, Minsk, 2008.

ed. Mishatkina Tatiana, and Melnov, Sergey, Environmental Ethics from A to Z, Minsk, 2008.
ed. Mishatkina, Tatiana, Practical Guide in Bio-Ethics and Environmental Ethics, Minsk, 2011.
ed. Mishatkina Tatiana, and Melnov, Sergey, Environmental Ethics: Glossary of a Young Ecologist, Minsk, 2011.
ed. Mishatkina Tatiana, and Melnov, Sergey, Social Bio-Ethics: Through the Lens of Global Bioethics, Minsk, 2018.
Petritsky, Vladimir, »Ecologisation of Morals and Ethics«, Philosophical Studies, No. 4, 1990, pp. 103–106;
Potter II, Van Rensselaer, Bioethics: Bridge to the Future (Prentice-Hall biological science series), Prentice-Hall, 1971.
Schweitzer, Albert, »Reverence for Life«, in Albert Schweitzer, ed. Guseynov, Abdusalam and Seleznev, Mikhail, Moscow, 1992, pp. 522–545.
ed. Tsyrdya, Teodor, Strategy of Survival in the Context of Bioethics, Anthropology, Philosophy, and Medicine, Kishinev, 2018.
Tulchinsky, Grigory, »Contemporary Humanitarian Paradigm«, in: Philosophical Century, ed. Tulchinsky, Grigory, Vol. 21, St Petersburg, 2002.
Vasilievene Nijole, »Environmental Ethics as Means of Goals Operationalisation for Sustainable Development«, in: *Environmental Ethics. The Power of Ethics for Sustainable Development,* ed. Vasiljeviene, Nijole and Jurciukonyte, Agne, Vilnius, 2010, pp. 289–329.

Ecological Civilization and Ecological Aesthetics in China: An Overview

Introduction

Since the beginning of the industrial revolution in the eighteenth century, the impact of human activities on the global ecosystem has been increasing dramatically. One of the direct consequences of the industrial civilization is the global ecological crisis. For the sustainable and healthy survival of the bio-community, including humankind, a new alterative type of civilization should be proposed and promoted globally. In Chinese, such an ecological civilization (i.e. eco-civilization) can be called ›*shengtai wenming*‹ (生态文明). In 2007, ecological civilization was proposed by Hu Jintao, general secretary of the Central Committee of the Communist Party of China (CPC), in his report to the 17th National Congress of the CPC (Lu 2019: 105)[1]. Since then, China takes the construction of eco-civilization as the national policy. Under this national atmosphere, ecological aesthetics (i.e. ecoaesthetics or eco-aesthetics) becomes one of the hot academic topics in China. Some scholars pay special attention to the relationship between ecological civilization and ecological aesthetics.

This paper firstly introduces the conception of ecological civilization in China and its international response. Then, the second section of the paper introduces ecological aesthetics before 2005. With this as the background, the third section focuses on ecological aesthetics promoted by ecological civilization in China and its unique contributions. In conclusion, the paper talks about the future of ecoaesthetics within the whole project of constructing eco-civilization.

1 The Conception of Ecological Civilization in China

The German academic, political scientist and researcher Iring Fetscher used the phrase ›ecological civilization‹ in his 1978 paper entitled

›Conditions for the Survival of Humanity: on the Dialectics of Progress‹. Fetscher (1978) argues that we need to change our present industrial civilization radically and to establish an ecologically balanced »alternative civilization«, which he calls »ecological civilization«. Briefly speaking, Fetscher's intention is to combine the problem of Marxism and ecology so as to reflect the problematic economic development optimism from the perspective of an ecological consciousness.

In China, the term ›ecological civilization‹ appeared in the 1980s in the academic domain and was then borrowed by political discourse. Ye Qianji proposed the ›construction of ecological civilization‹ in The National Ecological Agriculture Symposium in 1987 (Lu 2019: 106). It was taken up, developed, and vigorously and successfully promoted, most importantly by Pan Yue, the Vice-minister of China's State Environmental Protection Administration at that time. In 2007, the term was used by Hu Jintao, general secretary of the Central Committee of the Communist Party of China (CPC), in his report to the 17th National Congress of the CPC. Since then, official policy proposes that China should build an ecological civilization, creating more sustainable relations between production, consumption, distribution, and economic growth. Ecological civilization was incorporated into the official discourse as a new model of growth to replace the old unsustainable model. Rather than emphasizing economic construction as the core of development, as it did in the past, China had come to realize that development, should it be sustainable, must entail a list of elements including the right relationship between man and nature. It is not hard to see that there is a great change from the conception of ›sustainable development‹ to ›ecological civilization‹, which is from economic concern to philosophical rethinking. Ecological civilization was listed along with economic, political, cultural and social progress as one of the five goals in the country's overall development plan at the 18th National Congress of the Communist Party of China in 2012, when Xi Jinping was elected general secretary of the CPC Central Committee. The key tenets of ecological civilization include the need to respect, protect and adapt to nature, a commitment to resource conservation, environmental restoration and protection, recycling, low-carbon use, and sustainable development (Lu 2019: 112).

In Xi Jinping's report to the 19th CPC Congress in 2017, the construction of ecological civilization was emphasized dramatically. In the first part of the report, Xi stated approvingly that China had made much progress on ecological civilization since the last Congress. Notably, the report used terms like ›beautiful China‹ along with ›rich and strong, democratic, civilized, harmonious‹ to describe the goals for China's future. For the Chinese people, ›beautiful China‹ means ›green China‹, a China with a healthy ecological environment. The third part of the report concerned basic strategy. ›Greening‹ was listed along with ›creative, coordinated, open and sharing‹ as the five basic concepts for China's development. In the fourth part of the report, Xi stated that there were three great challenges that China faced: pollution, financial risks, and poverty relief. As the title ›Speeding reform of ecological civilization system and building beautiful China‹ indicates, part four of the report is about ecological civilization. According to the report, humans and nature are a life community. Humans must respect nature, follow nature, and protect nature. Only by following the principles of nature will humans avoid taking the wrong way in receiving nature's gifts. Part twelve of the report was on ›A Community of Shared Future for Mankind‹. It said that humans are facing many common threats, which no country can deal with alone. Xi called on all the peoples in the world to build a human community of shared future and construct a world with lasting peace, safety for all, shared prosperity, openness, inclusion, cleanliness, and beauty (Xi 2017: 588–601).

During the past 7 years, President Xi Jinping has often thematized the construction of ecological civilization. For example:

> The country should protect the environment like one protects his eyes and treat the environment like one treats his life.
>
> Lucid waters and lush mountains are as valuable as gold and silver.
>
> Mountains, water, forest, farmland and lakes are all parts of a common living community […] the administration of their use and ecological rehabilitation must be in accordance with the law of nature (CPC Central Committee 2017: 8, 21, 47).

Although there is not any clear and widely accepted definition of ecological civilization within China, the significance of the notion can be summed up as follows: (1) to reflect upon the forces driving the whole of humanity to ecocide[2]; (2) to reflect upon human future within the framework of three historical stages of human civilization: agricultural civilization, industrial civilization, and ecological civilization; (3) as the successor to industrial civilization, the goal of ecological civilization is very helpful in healing ecological destruction and catastrophe caused by industrial civilization; (4) ecological civilization enables us to clarify the meaning of civilization, to bring into focus the decadent barbarism of late capitalism[3], and to envisage more clearly its transcendence; (5) the notion provides a new vision of the future that can capture people's imagination and mobilize them for the required social transformation, say, from modern capitalism to eco-socialism[4].

The on-going project of constructing ecological civilization in China has attracted an international response. The phrase ›ecological civilization‹ wiggled its way into the lexicon of the international community at the United Nations Environment Programme Governing Council meeting in 2013 in Nairobi. The Council appreciated the Chinese delegation's promotion of ecological civilization as an effort to achieve sustainable development in light of national conditions. This delegation reassured its counterparts that ecological civilization is a more comprehensive expression of sustainable development. Australian scholar Arran Gare has published some outstanding work on the nature and philosophical foundation of ecological civilization in China. His book in 2017 expresses his basic points and statements, which can be summed up as follows: (1) The deficient assumptions of modernity led to the civilization of modernity, which must be replaced by ecological civilization, i.e., an ecologically sustainable civilization, a civilization based on process metaphysics and a fusion of science and the humanities in human ecology; (2) Civilizations are characterized by deep assumptions about the nature of the world and the place of humanity within it, and thereby the ultimate ends worth striving for; (3) Ecological civilization will involve a transformation in deep assumptions, ways of thinking, and ultimate aspirations; (4) What is required is the means to achieve this integration in order to mobilize humanity to redefine its place in the cosmos and its destiny, and then on this foundation, to create a society that augments life and the current regime of the global ecosystem. This will require a reorientation of philosophy to overcome fragmentation and replace the defective philosophical assumptions that have dominated modernity, including mainstream science.[5]

2 Ecological Aesthetics before 2005

It can be certainly said that there is a close connection between civilization and aesthetics. As a very complex term, civilization may mean an advanced state of human society; the act or process of civilizing, as by bringing out of a savage, uneducated, or unrefined state, or of being civilized; or cultural refinement, refinement of thought and cultural appreciation, and so on. Aesthetic activities always play important roles in the process of civilizing and cultural refinement, which can be mostly represented by the idea of ›aesthetic education‹. So, as a modern discipline, aesthetics lies at the heart of civilization. However, the conceptions of ecological aesthetics and ecological civilization emerged separately and developed independently until 2005. Let's firstly take a brief look at ecological aesthetics before 2005.

American ecologist Aldo Leopold's ›Conservation Esthetic‹ might be viewed as the early form of ecological aesthetics, because it takes the keyword ›community‹ borrowed from ecology as its theoretical foundation. In his 1949 book *A Sand County Almanac*, we can find the following influential maxim: ›A thing is right when it tends to preserve the integrity, stability, and beauty of the biotic community. It is wrong when it tends otherwise.‹[6] It is another American scholar named Joseph Meeker who firstly proposed the phrase of ›ecological esthetic‹. In his 1972 paper ›Notes Toward an Ecological Esthetic‹, Meeker's argument starts with the reflection of western aesthetic theories. He declares that since Plato western aesthetics has always been dominated by the great ›art versus nature‹ debate. Traditionally, aesthetic theory emphasized the separation of artistic from natural creation and assumed that art was the ›higher‹ or ›spiritualized‹ product of the human soul and ought not to be confused with the ›lower‹ or ›animal‹ world of biology. For Meeker, no matter we regard art as ›unnatural‹ product or as man's spiritual transcendence over nature, both ideas distort the relationship of nature and art. With this analysis as background, Meeker asserts that Darwin's evolutionary theory shows the evolution processes of living creatures and indicates that traditional anthropocentric thinking has overestimated human spirituality and underestimated biological complexity. From the 19th century, philosophers began to re-examine the closeness of biology and humanity and began to re-evaluate aesthetic theory in the light of new biological knowledge. Under such kind of research thinking, Meeker asserts that »aesthetic theory may be more successful in defining beauty when it has incorporated some of the conceptions of nature and its processes which have been formulated by contemporary biologists and ecologists« (Meeker 1974: 124–125).

Briefly put, the strategy of Meeker's ecological aesthetics is as follows: taking Darwin's theory of biological evolution as a theoretical foundation and laying emphasis on humankind's biological nature, reflecting upon and reconstructing aesthetic theory in light of contemporary biological and ecological knowledge. Meeker also criticizes the intellectual bias between natural scientists and humanists and advocates crossing disciplinary boundaries between science and humanity. It is especially meaningful that Meeker points out that ecology is a potent new model of reality which offers good opportunity to reconcile the products of humanistic and scientific investigation. He concludes that »ecology demonstrates the inter-penetrability of man and the natural environment« (Meeker 1974: 136). This conclusion indicates clearly the theoretical orientation of Meeker's ecological aesthetics: drawing lessons from ecological knowledge fully and grounding the aesthetic research on the science of ecology.

Meeker's research is mainly a theoretical one. In contrast, Paul Gobster (1999), American Research Landscape Architect with the USDA Forest Service's Northern Research Station in Chicago, develops what he calls ›An Ecological Aesthetic for Forest Landscape Management‹. His research broadly addresses the question of how can we design, plan for, and manage landscapes to sustain mutually beneficial relationships between people and ecological systems. The context for much of his work has been park and forest landscapes in urban and wildland settings, with the focus on three interrelated topic areas: perception and experience of landscapes, meanings of nature, and access and equity issues. Among the three topics, the first one is closely connected to the aesthetic field, because it aims at the question of how people perceive and experience parks and forests. The second one is actually a philosophy of nature and its practical application, trying to answer the question of how nature is understood and valued by different individuals and stakeholder groups, and how these values can be incorporated into landscape restoration and management. Gobster (1995) seems to intend to develop Leopold's ecological aesthetics so as to integrate aesthetic and biodiversity values and to resolve conflicts between them.[7] The attractive feature of Gobster's work is his effort of what he calls »Linking Ecology, Sustainability and Aesthetics«

(Gobster 2000),[8] or his main concern for »What does Aesthetics have to do with Ecology« (Gobster 2007)?[9]

Limited by the conditions of international communication in 1990s, Chinese scholars did not have enough access to ecological aesthetics in the west. To some extent, they developed their own theories of ecological aesthetics without reference to the literature mentioned above. The landmark of this special field in China is ›The First Nationwide Symposium of Ecological Aesthetic in China‹ in 2001, the formal beginning of the new century. Both Zeng Fanren and Cheng Xiangzhan participated in the conference and began their works on ecoaesthetics. Zeng Fanren's main idea of ›ecological existential aesthetic conception‹ is found in his conference paper in 2001, which is entitled »Ecological Aesthetics: A New Aesthetic Conception of Ecological Existence in the Post-modern Context«, which was published in the following year. The paper asserts that ecological aesthetics first emerged in a postmodern context. It shares a close relationship with both the continuing crisis of environmental degradation set on by pollution, nuclear threat, as well as developments in the field of deep ecology. Specifically, ecological aesthetics is an aesthetic perspective within ecological existentialism which holds significant implications with regards to shifts in perspectives of modern aesthetics, the evolution of eco-literature and the restoration of traditional Chinese ecological wisdom. Issues such as the proper conceptualization of the discipline itself, appropriate measures in addressing the re-enchantment of the world, establishing an intellectual position on modernity as well as science and technology, and its relationship with the field of practical aesthetics are but several among a whole host of matters awaiting resolve in the field of ecological aesthetics.[10] Based on this paper, Zeng developed his theory very rapidly and published his first book in 2003 entitled *Collected Papers on Aesthetics of Ecological Existence*. The core idea of the book is a phrase, ›ecological existence‹, which is an ecological interpretation of Martin Heidegger's fundamental ontology.[11]

3 Ecological Aesthetics and Ecological Civilization in China

Based on the above, the main trend of ecological aesthetics before 2005 can be summarized as ›Linking aesthetics with ecology‹. The story changed dramatically since 2005, which can be called ›Linking aesthetics with ecological civilization‹.

In 2005, the concept of ecological civilization was taken as a new vision for research on ecological aesthetics. This change was visible in the conference ›An International Conference on Aesthetics and Literature in the Vision of Contemporary Ecological Civilization‹ hosted by Shandong University in 2005. There were more than 180 participants from the United Kingdom, Norway, Finland, Holland, South Korea, Japan and China. The collection of the conference papers was formally published in 2006 entitled *Man and Nature: Aesthetics and Literature in the Vision of Contemporary Ecological Civilization.* Zeng Fanren submitted his conference paper entitled »The Notion of Ecological Aesthetics in the Vision of Contemporary Ecological Civilization«, which is based on an article entitled ›On Ecological Civilization‹ published in *Guangming Daily,* April 30, 2004. The newspaper article asserts that human civilization is in the process of transforming into ecological civilization, which is a new form of civilization after industrial civilization. Borrowing some key ideas from the *Guangming Daily* article, Zeng asserts that the coming of ecological civilization raises a series of new questions for aesthetics, promoting the aesthetics in industrial civilization to transform in order to adapt to ecological civilization.[12] Following the 2005 conference theme, Zeng Fanren continued his inquiry into the relationship between ecological aesthetics and ecological civilization. He entitled his 2013 book as *The Exploration and Dialogue of Aesthetics in the Era of Ecological Civilization,* which is a collection of his related papers and articles in the new century. Based on the ecological nature of human existence, the book explores the way to unify the ecological, humanistic and aesthetic conceptions, the ideal of ecological aesthetics.[13]

Cheng Xiangzhan thinks that the notion of ecological civilization guides a new direction for the future of ecological aesthetics. Inspired by the ideas of respecting, protecting and adapting to nature promoted by the 18th National Congress of the CPC (2012), he defines ecological civilization as a ›civilization in harmony with nature‹: human agent as the creator and subject of civilization ought to respect nature ethically, to adapt to nature in the ways of production and living, and to protect nature in practical actions. According to his understanding, the significance of promoting the ecological civilization for the constructing of ecological aesthetics can be summed up in the following four points: 1) leading to explore ecological aesthetics with the focus on the relationship between aesthetic activity and ecological civilization; 2) to ecologicalize[14] the aesthetic subject from Cartesian subject to embodied subject; 3) to ecologicalize the aesthetic object, i.e., objects appeared in ecological awareness, so as to differ ecological aesthetics from nature aesthetics and environmental aesthetics; and 4) to put the theory of ecological aesthetics into the ecological practices such as urban and rural planning and design so as to build up beautiful China.[15]

With shared interests and visions, Zeng Fanren and Cheng Xiangzhan launched an academic dialogue titled ›The Construction of Aesthetics in the Age of Eco-civilization: A Dialogue on the Idea of Eco-civilization and Today's Transformation of Chinese Aesthetics‹. They assert that the emergence and flourishing of eco-aesthetics in China in the 21st century is a historical necessary response to the new age of eco-civilization. The transformation from industrial civilization to eco-civilization is not only a significant economic and social transformation, but also a transformation from modern western anthropocentrism to eco-humanism or eco-holism, and a transformation from modern aesthetics of subjectivity to aesthetics of ecological existence. Eco-aesthetics not only takes ecology as its base, but also takes eco-philosophy related to ecology, especially Heidegger's philosophy of ecological existence, as its base. Chinese eco-aesthetics borrows a lot of theoretical resources from western environmental aesthetics and is closely connected with the economic background of traditional Chinese agricultural society and the idea of ›the unity of nature and mankind‹ in traditional Chinese philosophy.[16]

4 Conclusion

It is well known that the global ecological crisis is threatening the very existence of humanity. Some scholars including myself assert that the most radical response to global ecological disaster is the creation of a new, ecological civilization. That is to say, an international transition to new eco-civilization is necessary for all human beings. The ongoing Chinese project of ›Advancing Ecological Civilization and Building a Beautiful China‹ is the significant social practice globally.

Aesthetics lies at the heart of civilization. Ecological aesthetics is the ecological transition of aesthetics moving towards ecological civilization. Both of them need eco-friendly deep assumptions and ideals behind them, which should be explored philosophically. The unique feature of ecological aesthetics in China is its close connection with ecological civilization, which means that the theoretical construction of ecological aesthetics is one intrinsic part of the whole project of constructing ecological civilization. They should encourage and promote each other. The future of human survival depends on the successful construction of ecological civilization, to which ecological aesthetics should contribute more positive power.

—Cheng Xiangzhan, Shandong University, China

Literature

Cheng Xiangzhan, »The Notion of Ecological Civilization and the Direction of Ecological Aesthetics«, *Diverse Review,* Vol. 2, 2014, pp. 17–19.

Fetscher, Iring, »Conditions for the Survival of Humanity: on the Dialectics of Progress«, *Universitas* 30 (1978), pp. 161–172.

Gare, Arran, *The Philosophical Foundations of Ecological Civilization: A Manifesto for the Future,* London and New York: 2017.

Lu Feng, *Ecological Civilization and Beautiful China,* Beijing, 2019.

Gobster, Paul, H., »Aldo Leopold's Ecological Esthetic: Integrating Esthetic and Biodiversity Values«, *Journal of Forestry,* Vol. 93, No. 2, 1995, pp. 6–10.

Gobster, Paul, H., »An Ecological Aesthetic for Forest Landscape Management«, *Landscape Journal,* Vol. 18, No. 1, 1999, pp. 54–64.

Gobster, Paul, H., »Foreword«, in: *Forests and Landscapes: Linking Ecology, Sustainability and Aesthetics,* ed. Sheppard, Stephen R. J. and Harshaw, Howard, (IUFRO Research Series 6), Oxfordshire, UK, 2000, pp. xxi-xxviii.

See Gobster, Paul, H., Nassauer, Joan I., Daniel, Terry C., and Fry, Gary, »The Shared Landscape: What does Aesthetics have to do with Ecology?«, *Landscape Ecology,* Vol. 22, No. 7, 2007, 959–972.

Leopold, Aldo, *A Sand County Almanac and Sketches Here and There,* New York, 1968.

Meeker, Joseph, W., *The Comedy of Survival: Studies in Literary Ecology,* New York, 1974 [Revised as Chapter 6; originally published as »Notes Toward an Ecological Esthetic«, *Canadian Fiction Magazine,* Vol. 2 1972, pp. 4–15].

ed. The Party Literature Research Office of the CPC Central Committee, *Selected Discourses about the Construction of Socialist Ecological Civilization by Xi Jinping,* Beijing, 2017.

Xi, Jingping, *The Governance of China,* Beijing, 2017.

Zeng Fanren, *Collected Papers on Aesthetics of Ecological Existence,* Chang Chun, 2003.

ed. Zeng Fanren, *Man and Nature: Aesthetics and Literature in the Vision of Contemporary Ecological Civilization*, Zheng Zhou, 2006.
Zeng Fanren, *The Exploration and Dialogue of Aesthetics in the Era of Ecological Civilization*, Ji'nan, 2013.
Zeng Fanren, »The Notion of Ecological Aesthetics in the Vision of Contemporary Ecological Civilization«, *Critical Theory* 1, No. 2, 2017, pp. 8–31.
Zeng Fanren and Cheng Xiangzhan, »The Construction of Aesthetics in the Age of Eco-civilization: A Dialogue on the Idea of Eco-civilization and Today's Transformation of Chinese Aesthetics«, *The Journal of Poyang Lake*, Vol. 3, 2014, pp. 100–114.

Notes

[1] Lu's book offers a brief historical narration of the development of the concept of ecological civilization. See Lu (2019: 105–113).
[2] The neologism ›ecocide‹ can be used to refer to any extensive destruction of the natural environment and disruption or loss of ecosystem(s) of a given territory to such an extent that the survival of the inhabitants of that territory is endangered.
[3] Late capitalism is a term coined by continental European socialists in the late 1930s that has come to refer to modern capitalism from World War II onward. Later capitalism refers to the historical epoch since 1940, including the post–World War II economic expansion called the golden age of capitalism.
[4] Ecosocialism is a vision of a transformed society in harmony with nature, and the development of practices that can attain it. It is directed toward alternatives to all socially and ecologically destructive systems, such as patriarchy, racism, homophobia and the fossil-fuel based economy. It is based on a perspective that regards other species and natural ecosystems as valuable in themselves and as partners in a common destiny.
[5] See Gare (2017).
[6] See Leopold (1968: 224–225).
[7] See Gobster (1995).
[8] See Gobster (2000: xxi-xxviii).
[9] See Gobster et al. (2007).
[10] The paper was translated into English and published in *Critical Theory* (2017). The English journal's chief editor is Zeng Jun of Shanghai University, and Cheng Xiangzhan was invited to act as the special chief editor of the volume.
[11] See Zeng (2003).
[12] See Zeng (2006).
[13] Zeng (2013).
[14] North American environmental education guru, David Orr, explains that all education is environmental education, so too all education can be ecologicalized—can help teach and reinforce ecological understandings.
[15] See Cheng (2014: 17–19).
[16] Zeng and Cheng (2014).

Rewilding and Neoliberal Territorialities after the Anthropocene: Cybernetic Modelling of the Oriental Stork as Critique

This chapter explores the example of the reintroduction of the oriental stork *(Ciconia boyciana)*, sometimes called the oriental white stork since it resembles the white stork *(Ciconia ciconia)*. It went extinct in Japan in 1971, while the species survives in continental East Asia as an endangered species (class I) with conservation groups estimating that between 1,000 and 2,499 remain in the wild. Over the last three decades, the Hyogo Prefectural University has developed an institution known as Hyogo Park of the Oriental White Stork to plan activities in the name of the reintroduction of the oriental storks from Russia. It is a multi-partied joint effort involving cooperation of community residents and businesses, local farmers, as well as various local and national government entities, including those dealing with tourism and culture, not to mention the participation of species cultivated in the stork food chain.[1] It is this interdependency that forms the basis for analysis of territoriality and extinction in light of Deleuze and Guattari's posthuman cybernetics interdependencies, which I hope to show helps clarify misconceptions concerning extinction in existing conservation discourse and practice in Japan. In the course of this investigation, it is suggested that the concept of the Anthropocene is a misnomer, that some form of post-Anthropocene era should be introduced to account for the cybernetic interdependencies in the due course of the human generation of self-displacing posthuman decentring. It is suggested that conservation requires cold post-anthropocentric modelling to account for its responsibilities that exceed modelling based on the Anthropocene as an era of human ascendency.

Before the Fukushima Disaster, Japanese nuclear regulatory agencies and corporations ignored warnings and sustained their established method of relying on cover-ups of known dangers even as the Disaster unfolded. Though one would expect such behaviour to change given the rarity of known meltdowns in the world, it forms a benchmark now for understanding the priority placed on profits (a hallmark of neoliberalism) over biological sustainability of life systems including humans. Subsequent indifference to human and nonhuman victims was evident, as if the disaster only reinforced a brash sense of blindness to material conditions and a retrenching of state authority and priority placed on corporate profits. In relation to conservationism in neoliberal Japan at this time, one must ask, to what extent have ecologically minded Japanese become ineffective or similarly misled in the current neoliberal context, which is partially responsible for placing power industry profits over the sustainability of life in Japan.[2] To answer this question, relevant to conservationism in Japan, demands that attention be drawn to historical state-driven enterprises as well as contemporary neoliberal withdrawals of state responsibility for endangered species (not to mention the growing precariousness of people's lives)[3] and extinction in the context of the oriental stork. People, livestock, wildlife and land are expendable, but capital invested in nuclear power infrastructure is not; this is the lesson of Fukushima. Similar issues have been addressed in critiques of Anglo-American conservation efforts, as when Lorimer (2015) situates the role of neoliberalism in privatizing and marketing charismatic flagship species within its consumer-oriented modelling of society and ecosystems as mere halfway measures (Lorimer 2015). Lorimer's suggestion of a matrix of ecological charisma in effect outlines human-nonhuman affects as forms of communication. He creates, echoing Deleuze and Guattari's uses of the Spinozan sense of affect as generalised force (not personal emotion), a capacity to affect and be affected—a Nietzschean differentiation of passive (or reactive) affect and active affective engagement that alters contextual conditions. In conservation, this distinction may be extended not simply to inter-human politics but human-nonhuman as well as nonhuman-nonhuman relationality among species.

If, as Nigel Thrift (2008) summarises, »in the world of Spinoza and Deleuze, affect is the capacity of interaction that is akin to a natural force of emergence« (Thrift 2008: 182), then affect may include not only human communication but any species or inter-species signalling. With respect to the storks, their affective relations—conscious or autonomic—require not human presence but indeed distance to reproduce and flourish. Conservation focusing on viewing the storks may build human interest in them but be perceived as threatening to storks and not help them at the level of establishing conditions for affective emergence as a species in a sustainable ecosystem. Kathleen Stewart (2007), very helpfully in the context of conservation discourse, defines affects as »a kind of contact zone where the overdeterminations of circulations, events, conditions, technologies, and flows of power literally take place«. They are »at once abstract and concrete« and »more directly compelling than ideologies, as well as more fractious, multiplicitous, and unpredictable than symbolic meanings« (Stewart 2007: 3).

Deleuze and Guattari's territorial metaphors may be used to frame species narratives of extinction and recovery. In particular, they ask how capitalist forms of culture and exchange—given their schizophrenic array of material-abstract assemblages—may responsibly situate species decline and concomitant questions of reintroduction and recovery. In addition to de-/re-territorialization, Guattari's concepts of transversality and ecosophy will be critically assessed in light of parallels in stork conservation discourse. Then, one may see how Deleuze and Guattari's critical metaphors deconstruct and redefine the givens of neoliberal Japan and its compromising approaches to reintroducing locally extinct charismatic megafauna: systems of nationalism, tourism, pollution, transportation and urbanization that continue to undermine ecosystems in the first place. In the end, extant territorialities will be shown to highlight the ongoing counterpoint between pragmatics in acknowledging stork needs and conservation values that are sometime reliant on obfuscating abstractions such as a Japanese oneness with nature.

1 The Japanese Reintroduction Project: Oriental Stork as Automata of Hope

Prominent narratives offered by Toyooka City and Hyogo Prefecture documents conflate ancient times with an ideal to be recovered in time future, with few attempts to mediate time in-between: the time when oriental stork apparently just vanished. Yoshito Ohsako (2011) writes: »Japanese people lived in harmony with the storks in agriculture, forestry, tourism, entertainment, local knowledge and so on in the past. The storks were considered as a part of the scenery and customs for the people«. This text, submitted for publication through Hyogo Prefectural Government channels, plays on clichés of Japanese living in harmony with the storks in a timeless past that anachronistically includes contemporary institutions (agriculture, forestry, tourism), confusing emotive ideological fantasy with contemporary conservation efforts.

The fate of the Oriental stork would seem to hinge on two factors. First, the *scale* of wetland habitats necessary for successful breeding and repopulation of a species reduced to an estimated two to three thousand birds inclusive of all its known territories in East Asia. Second, various forms of cooperation with humans at all levels so as to not only refrain from killing the birds but to proactively engage in cultivating toxin-free wetlands that allow stork prey to form supportive ecosystems within which the bird (not to mention other species) may once again thrive. After all, a rice field grown with pesticides and fertilisers may look green but could be as helpful as a parking lot to the stork if biodiversity is inhibited. In Japan, governments sponsor programs to flood unused farmlands so as to create wetland ecosystems where storks can consume frogs and dragonflies; however, recent research shows that if repopulation is the aim and expanding wetland or rice fields into new territories for Oriental stork is the means, the scale must be larger and contiguous in order for successful reproduction to be expected.

The very modern ideology and language of the fiction of oneness of a Japanese people, tied to oneness with nature, suddenly grows correspondingly distant from an archipelago colonised by networks of trains and highways, urban sprawl, and chemical overuse destroying species in rural farming areas. Yet, as will be shown, some Japanese conservation professors and advocates cling to such ideologies, apparently not realizing their counterproductive reifications of wartime nationalist ideology (not ethnic or inherited exceptionality) that they have embraced.

Japanese (postcolonial) war machinic affective engagement with the world can be shown to constitute a conservative falling back on a simplistic ›nature-Japan‹ fusion, that is more accurately a mental subtracting of things western from the category of being Japan and being Japanese so as to produce a nature-Japan (my coinage) utopic fantasy. Deleuze and Guattari's conception of a »war machine« (Deleuze and Guattari 1987, Ch. 12) explores precisely such spatio-temporal distortions, the sort that occur when engaging a movement *toward* war, whether engaged or not, as in the pseudo-war of competition during the period of high-growth capital in 1960s-1980s Japan, and by analogy the neoliberalisation and austerity measures to follow. However the nature-Japan discourse and practices may exhibit such war machinic affects and family resemblances. In terms of territoriality, the innovative means of sharing paddies with the stork suggests not simply a return to past practices but, quite to the contrary, an active posthuman conservation effort that modestly stands out against the European and North American nature reserves cultures, which are characterised by stark demarcations of areas absented of humans (Lorimer 2015: 163–64). Stork conservation efforts in Japan have avoided the »territorial trap« (Lorimer 2015:164) of conserving species and ecosystems by way of mutually exclusive human-nonhuman spaces. Still, within the context of extreme neoliberal protections of private (including corporate) interests, a more flexible »fluid topography for wildlife« (Lorimer 2015: 176) as a means of connecting stork populations with other species' needs remains limited. Though academics in conservation understand the necessity of »nonhuman mobilities« (Lorimer 2015: 177) and the need to cultivate many species, the overall planning methods stress limitation to the ›doable‹ within the existing demands of human cultures (farming, tourism, etc.) which inhibit and police the nonhuman despite the best of intentions. Part of the inherent limitation stems from being blinded by nationalist attachments for grounding conservation thought in irrational ›the west‹ versus Japan binaries.

Although conservation in part is about *moving* citizens into action, the actions that really matter are decisions made by farmers and government agencies to regulate practices that pollute or dry out wetlands and paddies or lead to the loss of tall trees or artificial poles required by oriental storks for nesting. *Conservation that focuses on big ideas risks drawing attention from the cybernetic entanglements that matter and that require some variety of posthuman compromise with the needs of other species.* One may at this point note how Claire Colebrook (2014), in a sweeping exploration of theory's relation to extinction, misreads posthumanism in this respect. Posthumanism may be understood as a discourse predicated on a revaluation of human hubris in human power or territorial relations with other species and their needs in various symbolic and material manifestations that recognise interdependencies.[4] It by no means »begins from an already integrated, dynamic and connected world« (Colebrook 2014: 20), but rather, following N. Katherine Hayles (1999), is a critical intervention itself always in a state of becoming. It is not a given. After all, it is ›post-‹ in the sense of ›no longer only‹ human, not simply in the sense of being absolutely ›no longer‹ human. Colebrook seems to see posthumanism as abandoning all semblance of human agency, an extreme position not borne out in the literature either in posthumanism or conservationism, and certainly not in Deleuze and Guattari's philosophy of immanence that builds on territorial site-based compromises found in species interaction, not in self-effacement.[5] It is here, in their virtual-material coordinates (that in some ways replaces Cartesian coordinates of thought and being) that Japanese nature nationalism displays its wishful humanist entrenchment. Nature colonization only seems to be an inoffensive register for grounding nationalist projects, as it seems on the surface innocuous and to dovetail with the general ecological crisis precipitated by anthropogenic climate change.

The Investigative Committee on the Reintroduction of Oriental White Storks (2014) reports key successes attributed to local activities, including the increase of the wetlands along the Maruyama River from 82 hectares in 2004 to 127 hectares in 2014, approaching 1930 levels when the stork population was 100 (Investigative Committee 2014: 4, 12). They also document the adoption of organic farming as ›Stork Friendly Farming‹ (Investigative Committee 2014: 4) and various affiliated products such as the stork-branded rice labels associated with lower levels of pesticides or organic grains. But, the report also reifies abstractions derivative of Japanese nationalist sentiment, including the clichés that Japanese love nature (Asquith, and Kalland 1997: Ch. 1) and have a conventional longing for their rural hometown (even when most Japanese live and are raised in urban environments). These abstractions are presented repeatedly in the report by the Investigative Committee, a group composed primarily of university professors, who emphasise »an analysis focusing on ›sympathy‹ as the axis of the evaluation« (Investigative Committee 2014: 6). For instance, one text box titled ›Points of sympathy for seeking a better life‹ contains three numbered points: »Love of living creatures = Biophilia (Love of nature symbolized by Oriental white storks«, »Love of the hometown = Topophilia (love of a place »Toyooka«)«, and »Desire for a stable life« (Investigative Committee 2014: 6). These slogans are wishful thinking, perhaps not pertinent to the actual conditions in Toyooka, characterised by farmers abandoning their fields and residents relocating to urban areas. On the other hand, the report also emphasises more practical aims, such as raising the »level of recognition of the development of local communities living in co-existence with Oriental white storks« (Investigative Committee 2014: 11).

In a graphic rendering of Hyogo Toyooka Model by the same group of scholars assessing conservation efforts (fig. 1), despite the detail and complexity, one can see a process and relationality that promotes binary labelling in lieu of more intricate cybernetic engagement with posthuman force-struggles at play in situ. As this is primarily a local effort in Hyogo City and Prefecture, putting aside national government indifference today—contrasting with the roles of war-mobilization and national economic development in the decimation of stork populations—what appears in this graphic representation *seems* to be along the lines of Guattari's ecosophic (ecological philosophical) modelling of cooperative and mind-changing relationality that may proactively deconstruct capitalists' assumptions that ecosystems should serve profit-extraction processes and, to the contrary, transform human practices into more other-oriented and dynamically engaged ones. However, the capitulation to a neoliberalisation of ecotourist fixation on stork imagery (note the repeated use of the stork as »symbol« in a »process of awareness« and »a future image« and »visualization«) and affective production may endanger the well-being of storks if this symbolic chain of spectacular and fantastical imagery were ever to fail, to lose human attention. The use of the oriental stork to attract tourism and sell ›Flying Stork‹ brand rice suggests a reduction of the stork to a spectacle that in effect would contribute to passive reception of a purely emotive affectivity: a static image used in public relations campaigns and travel brochures as well as enforcing a ridiculous ecological engagement under the cartoon image of oriental stork. Then, storks again might suffer the same fate of extermination in the name of famine, war or economic competition (requiring, for instance, the use of pesticides that would destroy the stork food chain).

2 Temporality and Territoriality

Though the literature on the demise of the Oriental stork population plays down the role of militarism and war sacrifice in the decimation of the stork as an existential threat to wartime Japanese—as the bird has the habit of trampling rice plants—emphasis in terms of both activities and conceptualization of the reintroduction project is on

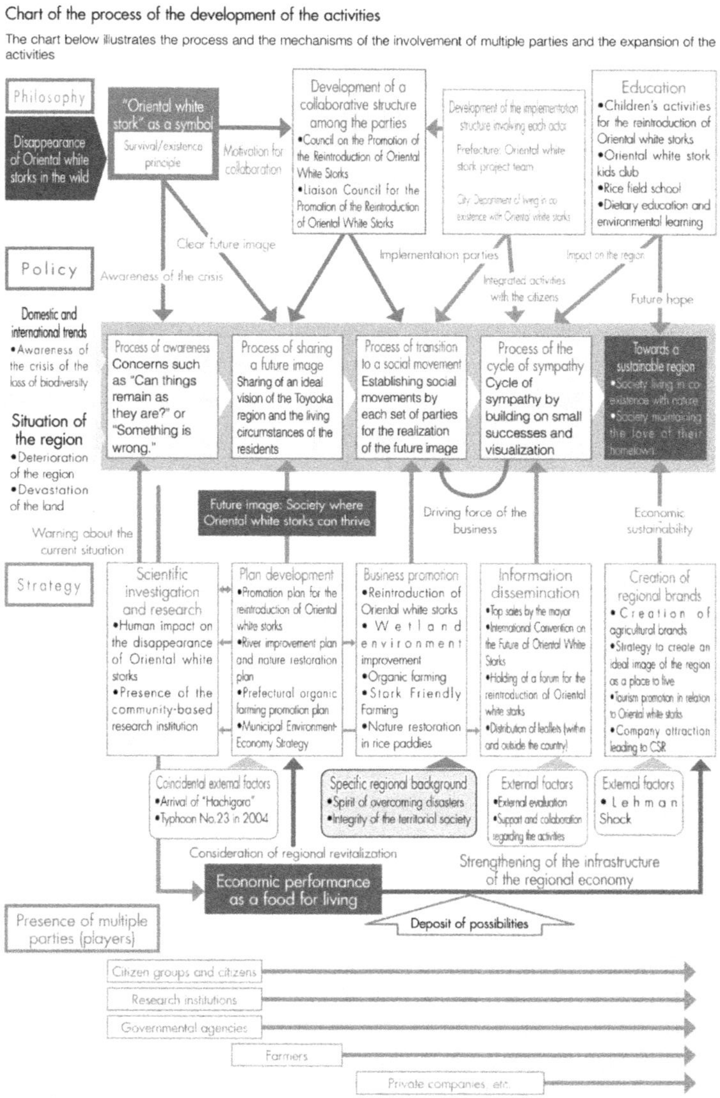

Figure 1. Excerpt (upper half) from *Assessment of the extension of activities for the reintroduction of Oriental white storks [summary]: The ›Hyogo Toyooka Model‹, which promotes local communities living in co-existence with Oriental white storks.* (Investigative Committee on the Reintroduction of Oriental White Storks 2014: 23).

altering human practices to make possible a transformation of the ecosystem as a whole. Oddly, it seems that the only untouched given of nature is the ecosystem of neoliberal capital itself. All else involving the storks is structured and planned as compromises. Thus one might better situate the reintroduction project as a manifestation of the reterritorialisation of the Capitalocene (Moore 2016).[6] Since the Anthropocene frames an age of human dominance, indeed the Capitalocene more precisely frames this less as an essential given manifest in the rise of humanity than as a particularly dangerous system within humanity's ever-expanding exploitation and occupation of the raw materials and biosphere of planet Earth.

As Christian Schwägerl writes (2014: xi), »Sustainability ideas come prepackaged with a set of imperatives. The Anthropocene idea works differently, but in a complementary way. It exposes us all and asks for responsibility«. In other words, naming an era of human domination need not consolidate a sense of capitulation to the inertia of a failed status quo; it could and should incite a renewed sense of determination and engagement in vigilant activities aimed at creative change. In other words, while planners and assessors of the plan highlight the laudable »Presence of multiple parties (players)« within the »Hyogo Toyooka Model« (see fig. 1), such abstract flow charts can be dismissed as bureaucratic simplification, reducing unknown sites of interaction and becoming to a capitalist model of investment and returns propelled and threatened by various contingencies. Core needs of the storks are not foregrounded as absolutes, while human institutions are.

3 Rewilding

Rewilding is not the same as sustainability. The former may imply a teleological movement and a return to an idealised period of an ecosystem in a given territory, say when the stork numbers were at a known peak. Sustainability maintains a more sensitive (and ecosophical) focus on the malleability of territories and actors themselves in dynamic relations within constantly changing situations. We may, as humans, a species capable of reflection and adaptation, situate the storks both in the Anthropocene and, in Guattari's sense, as part of the »artificiality« of the assemblages we always already inhabit. In Guattari's notes to *The Anti-Oedipus Papers*, he moves away from both a Lacanian absolute of the guiding »petit objet a« (Guattari 2006: 152), a remnant of the traumatic real, and the capitalist reification of value in symbolic language—which he associates with »totalization [of signifying practices] cut off from all praxis« (Guattari 2006: 87). It should be pointed out, however, that Guattari sees »molar bodies« as subject to group practices while »molar alliances« of groups (under signs and rules) may suppress freer »molecular filiation«. In the realm of conservation, note how molar alliances and molecular filiation are challenged by the local Hyogo government website (and presumable locally-distributed) pamphlets that emphasise the importance of emotional shifts that correspond with behaviour shifts, both of which would assist in the survival of the oriental stork. Thus, the conservation efforts lose focus on the application of research findings and therein lose focus on the *needs* of the oriental stork. Instead, in effect, focus on the *feeling* of doing something good for others as other is underscored. This is not to say filiation is nil, but just that it is overshadowed by »molar« group practices in concert with the form that oppresses the oriental stork: capitalist and other human demands for territory and, more specifically, consumerist subject models of identity, rather than posthuman (inter-dependent) individuation[7] that would sustain a constant renegotiation of the human-stork territorial needs.

Guattari's *The Three Ecologies* expands his earlier emphasis on transversality in psychiatric counselling, removing the doctor as authority figure and letting social interaction develop spontaneously to ecosophy as an ›ethico-aesthetic‹ approach bringing into relation »social ecology, mental ecology and environmental ecology« (Guattari: 2000: 41). He locates in environmental problems »a certain incomprehension and fatalistic passivity [...] among both individuals and governments«, accepting »negative developments [...] without question«. Representative of his critique of capital flows, he asserts the need for praxis in these three ecologies—a position that is helpful in situating Japanese conservationists, who range from nationalist ›nature-Japan‹ advocates to pragmatists.

But, in addressing problems of ideology in Japan, one may further turn to Guattari (2015: 106) who focuses on the need for changing how people think:

> ecological consciousness ought not to be satisfied with worrying about environmental factors, such as atmospheric pollution, the predictable conse-

quences of global warming, and the extinction of animals; it also ought to bear upon ecological devastation in the social and mental domains. Without transforming mentalities and collective habits, there will only be ›remedial‹ measures taken concerning the material environment.

Thus Guattari is helpful in understanding the problem of creating conservation models that are both pragmatic in their effectiveness and capable of altering human mentalities, in effect altering practices and conceptual individuations so as to create sustainable ecosystems.

4 Cybernetics, Rewilding Fantasies, and the Territoriality of Performative Ontologies

Although *cybernetics* typically refers to computational systems as applied in explorations of human-machine affinities and barriers between organic and artificial life or intelligence, in this work it is used to refer more generally to the dynamism of cross-species communicative interaction or the need to see beyond the nominalization of cybernetic organisms into essentialised objects or subjects. This follows approaches first explored in second-order cybernetics, which focuses on communication modelling, not merely machine interfaces. Such cybernetics, as Andrew Pickering (2010: 380) points out, engages the framing of ontologies in terms of conjunctions of complex »systems—human, nonhuman, or both—that staged their own performative dances of agency« so that »one can read cybernetics as ontological theatre«. Then, if we understand the ecosystem of the oriental stork in Japan as always partially unknowable »exceedingly complex systems« in which humans may only engage in interactive »performative dances of agency and findings-out« (Pickering 2010: 380), cybernetics becomes a helpful field of attention for situating the de- and re-territorializations among species in specific contexts.[8]

Cybernetics helps clarify the territorializing relations between humans engaged in (or opposed to) conservation policies and the storks, as well as numerous other species and material conditions that factor into their shared systems. Whereas *rewilding* suggests an infinite nature, the communicative relationality of species in a cybernetic model maintains the possibility of the knowability of a local event, while sustaining as much complexity as is necessary. Humans and oriental storks form mutual dependencies: humans use storks to draw tourists and sell ›stork-friendly‹ (mostly organic) rice, while storks require that humans stop using pesticides in the paddies and keep them flooded as long as possible year around in order to subsist if not thrive. We know that the territorial criteria for its continued survival are very specific: The oriental stork is a »wetland-obligate species« that »nests on tall trees and artificial structures such as electricity pylons and feeds on fish and small animals in open, usually fresh water wetlands, and occasionally coastal tidal flats« (Liu et al. 2008: 292). These systemic requirements reflect divergent virtual codes immanently manifest separately in humans and stork. Abstract capital is produced so that human economic systems remain balanced from the perspective of existing human values while the stork's genetically coded needs are also met.

It is in the writings of Deleuze and Guattari that the impact of the cybernetics-inspired de-centering of subjects and exteriorization of knowledge and processing of autopoietic worlds become especially useful, for they situate machinic relationality between species. Their work—especially that in the period from *A Thousand Plateaus* through Guattari's work on ecosophy—provides numerous in-roads for rethinking human relations with other species such as the oriental stork. Although much of the work relies on discussion of liberating human desires, so that Guattari's psychiatric background in some ways seems to inhibit the realization of a rigorous inter-species model of communication and understanding, indeed even transversality—entrenched as it is in the anti-psychiatry movement model of mutually de-hierarchised sharing (talking and listening)—might be applied to develop a better understanding of human-stork relationality in the case of Japan.

In *A Thousand Plateaus,* Deleuze and Guattari explore various relations between understanding human-nonhuman territoriality. In addition to their more well-known preoccupation with biosemiotic symbiotic relationships between the orchid and wasp or fly and spider, outlined in terms of genetic codes and observable behaviour, they situate assemblages of a less harmonious sort. This other work proves helpful in understanding the precarious position in which endangered species are placed due to human behaviour. Such »combinations are neither genetic nor structural; they are interkingdoms, unnatural participations. That is the only way Nature operates—against itself. This is a far cry from filiative production or hereditary reproduction« (Deleuze and Guattari 1987: 242). Their work on territoriality helps highlight materialist accounts of force and territorial interaction of species in the blind spots of neoliberal societies. They write: »A territory borrows from all the milieus; it bites into them, seizes them bodily (although it remains vulnerable to intrusions). It is built from aspects or portions of milieus. [...] It is by essence marked by ›indexes‹, which may be components taken from any of the milieus: materials, organic products, skin or membrane states, energy sources, action-perception condensates« (Deleuze and Guattari1987: 314). They define territoriality not simply as where physical occupations of places and spaces are present, but as a multifarious arena in which claims to territoriality are presented in relation to other places and qualities that define them.

By suggesting that cybernetics may itself serve as a critical platform for analysing the state of conservation efforts, I mean to build on N. Katherine Hayles's (1999) discussion of incorporating practices that stand apart from discursive practices in the production of knowledge. When the storks were resettled from Russia into the Hyogo reserve park and surrounding citizens cooperated by changing their farming practices, reorganizing irrigation to maintain water in paddies for longer period and allow for species of fish and frog to flourish for the benefit of stork, through new human practices and new norms became incorporated knowledge. The irrelevance of the nature nationalism in these re-inscriptions of discursive knowledge—as these new practices elude »conscious view« and redefine »the boundaries within which consciousness takes place«—may be understood as »new technologies that affect how people use their bodies and experience space and time« (Hayles 1999: 205). In this sense, foregrounding Deleuze and Guattari's cybernetic roots[9] allows for an even more critical grasp of territoriality in Japanese conservation efforts, and allows us to see how Japanese nationalism, although marshalled to promote big projects that have even been incorporated in international United Nations programs—such as the Satoyama Initiative[10]—would distract from the systemic (autopoietic) needs of nonhuman target species to which attention is being paid. As Hayles (1999: 137) writes, »Each living system thus constructs its environment through the ›domain of interactions‹ made possible by its autopoietic organization«. By focusing less on overzealous grand narratives of ›oneness‹ with nature—far too general to be of any practical use—cybernetic functionality (assessed in relation to sustainability issues) and beneficial practices to incorporate make a difference due to pragmatics, not ideology. And this forms a critical difference that I argue may be an asset to conservation itself. Just how such a shift would be implemented is beyond the scope of this paper.

5 The Deterritorialising Clatter of the Oriental Stork

One aspect that Deleuze and Guattari mention does not fit the neat songbird territorializing explored in *A Thousand Plateaus;* however, it does shed light on the interrelation of charismatic species and human cultural *changes* in Japan. Rather than rest on one's laurels under the monolithic idea that all Japanese live naturally with nature (if only the west had not contaminated Japanese beauty),[11] one can see the ancient poetry canonizing one bird while omitting another as setting up an aesthetic hierarchy that contributed to the extinction of stork. While one can blame the west, it was Japanese participation in imperialism that deserves the actual credit for decimating stork populations; during its expansionist period, storks were sacrificed in order to guard rice production or due to the felling of vital tall trees to support the war effort. Moreover, western conservation movements were originally joined by Japan and not the reverse. The idea that the stork is worth saving itself owes a debt to a reversal of the non-aestheticising history of the oriental stork. For lack of a better word, its ›birdcall‹ is unbecoming, in comparison to a staple of Japanese poetry and art: the Japanese bush warbler *(Cettia diphone;* 鶯 *uguisu),* whose call smoothly and dramatically builds in volume (and with varied pitch, not shown) so as to endear itself to poets in imperial anthologies dating back at least to the early tenth century. The clattering of an oriental stork is a monotonous sound historically not lending itself to the realm of aesthetic allure and production; it has been largely ignored. Compare the modulation of their calls in figures 2 and 3.

In *The Three Ecologies* Guattari provides a sketch for how to disentangle human subjectivity from its central role as mediator of worlds while retaining a role for human responsibility by way of communicational entanglement across discourses, what he calls ›semiotic regimes‹ of the economic, the juridical, and the techno-scientific varieties (Guattari 2000: 48). Guattari argues that the binary worker-corporation relationship itself is captive to a fictive division parallel to the division of human and non-human, master and servant, while actually »the economic-ecological vectors of circula-

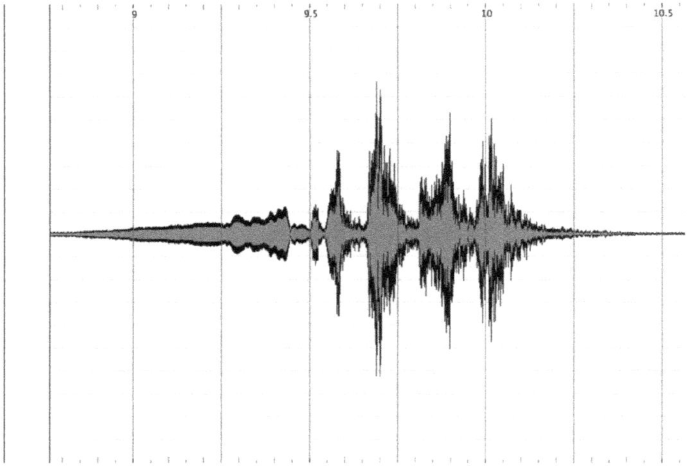

Figure 2. Call of the Japanese bush warbler.

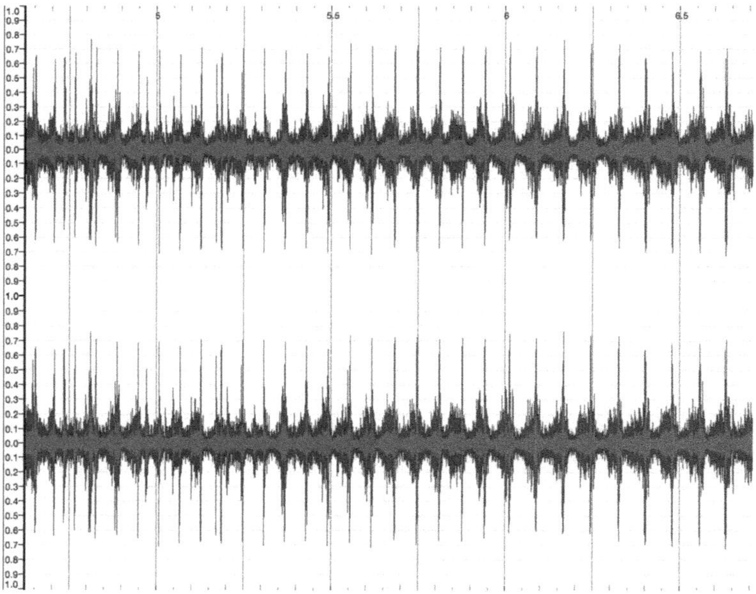

Figure 3. Clattering of the oriental stork.

tion, distribution, communication, supervision, and so on, are strictly situated on the same plane, from the point of view of the creation of surplus value, as labour that is directly incorporated into the production of material goods« (Guattari 2000: 49). Taking this model into the current eco-materialist turn toward acknowledging the alterity of non-human presences rather than a containing anthropocentric subjectivity, Guattari provides a rough model for disentangling the human economies and cultures not only from the *cul-de-sac* of human subject-production but from the needs of the nonhuman, which may then be distinguished and foregrounded. Thus endangered species and their precarious situations—often due to compromised territorial claims—may be mapped to the production of these blind spots in human awareness concomitant with human cultural and economic production.

6 Fallacies Concerning Japan and ›Nature‹

A favourite writer of the blame-the-west school is Watsuji Tetsurō, famous for his monograph *On Climate* (風土) that essentialises cultures as products of climates. In the Kyoto School (of philosophy) others engage in western treatments of Buddhist issues and develop hybrid discourses firmly within the frame of western philosophy but with the intention of overcoming it on ›Japanese‹ or more generally ›Asian‹ terms.[12] It is more the *intention* of leaving the west and ›returning to Japan‹ than the *embodiment* of such an aim. Similarly, there is a tendency, as Richard Reitan (2017) argues, for extremely conservative ecologists in Japanese to identify a problem »primarily in idealist rather than materialist terms, as a problem of ideas and beliefs rather than material conditions«.[13] The second overarching fallacy of this ecocriticism was perhaps influenced by various Kyoto School philosophers, who in turn were influenced by Hegel, Bergson, and various phenomenologists: the idea that human intentional constructs can perform superior forms of ›unity with nature‹ than contemporary discourse in posthumanism, which takes a more materialist approach to the inter-dependencies of species (including the human) in a more objective and yet engaged way.

Professors and directors of various museums and conservation as well as religious organizations contributing to the Biodiversity Network Japan's (pre-Fukushima) collection of essays, unconsciously (one would hope) include nationalist works reifying wartime clichés. For instance, Kunio Iwatsuki (2008: 10), Director, Museum of Nature and Human Activities, Hyogo and Professor emeritus, University of Tokyo, resorts to conflating economic and biological materialism, writing:

> Our extreme materialism often leads us to evaluate every material in its monetary value. In this way, it is rather difficult that all people understand the meaning of sustainability of our only Earth. We should be respectful to the donor of the materials to support our lives and thank nature as in traditional Japanese culture. On the basis of this idea, boys and girls can understand the worship of nature and will eagerly contribute to sustain the Earth.

The general fallacy may in part be attributed to confusing being primitive—a state in which *all* peoples are indigenous peoples and live off the land one way or another with minimal governmental structure—and being Japanese of a modern national state based on inculcated strict public order. Surely a modern person cannot claim to be acting on a Shinto or animistic instinct, since what is needed is not (or not only) a religious affect but a rational one capable of rethinking inter-species' needs and ecosystemic possibilities of complex and multiple systems.

As for the division of spiritual and material, one can point out, as Julia Adeney Thomas shows (2001: 181), that such putative spiritual and nature-bound Japanese culture »is quite plainly not a traditional notion. It differs markedly from the universalism of most Tokugawa and Meiji conceptions in claiming that there is a form of nature unique to Japan«. Thomas (2001: 181) shows how wartime apologists, both »scholars and bureaucrats [...] created this image [and] sought to root it in antiquity by scavenging the past for examples of Japanese devotion to nature« so as to »craft a convincing aura of continuity«, a »vision of a Japan dominated by age-old *Gemeinschaft (kyōdōtai)* intimacies« and »immemorial harmony with nature«. This critique puts into question whether legacies of wartime ideology can or should be martialled in conservation efforts. Can such a situating of an anachronistically retrofitted tradition, with respect to human-nature harmony, address the deeply individualistic ideology of neoliberalism prevalent in Japan today? How can bland appeals to abstractions such as spirit and general animistic practices do more than preserve the ruling neoliberal order? In light of Fukushima, one cannot simply ignore the business-as-usual lack of dissent. Material concerns would seem to matter more, not less, and nature merely become another territory to colonise. Discourses on Japanese race (a direct result of wartime propaganda) are even reflected in stork-reintroduction studies by way of its implicit emphasis on species lineage purity.[14]

To his great credit, Hidemichi Ootagaki (2008), Director of the Oriental White Stork and Human Coexistence Department, Toyooka City, in the same Biodiversity Network Japan pamphlet, by contrast refrains from entering into the fantasy of disentangling Japan from the west in nationalist restorations of pre-westernised Japanese views of nature, which are themselves anachronistic constructs. Nevertheless, his argument focuses on integration of stork needs with human economic needs, thus acceding to the neoliberal model and offering no critical resistance to its systemic antagonisms towards non-profitable nonhuman species needs. In other words, there is no vision inclusive of non-charismatic species conservation projects.

7 Conclusion: After the Anthropocene

The human encoding of a self-recognised species-specific transformation of the planet earth as the ›Anthropocene‹ has the misfortune of underscoring our differentiation from other species as if it were a given state of being preeminent rather than a precarious process of multidimensional entanglement. In fact, it might be more helpful to call it the ›post-Anthropocene‹ since at the very point of approaching such human mastery of the planet—through digital, archival, computational, chemical, nano, and genetic technologies that allow the rethinking and modification of matter itself—humans not only demonstrate control over other species but become subject to being controlled by our creations and transformations. The monstrosities released form our new ›nature‹, or cybernetic entanglements and dependencies for food, air, sustainability of ecosystems, and so on. That is, ecosystems are (except in controlled laboratory conditions) always subject to unforeseen inter-species as well as bio-chemical events. Multi-directional communication must define any serious manifestation of the Anthropocene; it behoves humans to watch, listen, measure, and learn, not merely to construct tolerable habits for a putative ›rewilding‹.

Such post-Anthropocene interdependency would undermine individuated relations between organisms, whether basing relationality on categorical definitions of functions or material controls (vectors of force). Thus, if any postnatural rewilding is to take place after the loss of the fiction of pristine nature, it must include us, but not only us. Rewilding in the post-Anthropocene might mean the dynamic application of all our technological prowess poised in an overhuman cultivation of sustainable ecological systems. Surely, we—at 7.5 billion—could not simply all retreat now to an indiscriminate wild to live in trees in some ›deep ecological‹ entrenchment. Clark (2015: 61) locates in *fire* a technological development »lighting a space against the dark« that suggests both a turning point, and point of no return, in its refusal to give up technological conveniences. It is at this juncture that one may consider how human actions have first led to the local extinction and then to the reintroduction of the oriental stork in Japan. At the cusp of the post-Anthropocene, with its *uncontrolled* dialectical entanglement among species, posthumans would subject themselves to the species they choose to conserve, operate following their *jouissance* so as to save some species and ignore the threatened status of others. Are not technologically sophisticated Japanese, by no means unified, already making their decisions, that the preservation of capital is of supreme importance, and species may be allocated according to economic feasibility? If so, it would partially explain the continued hunting of whales as well as the slaughter of porpoises and dolphins (McCurry, 2014) by some Japanese fisher persons who see them as competitors if not always as lucrative prey.

As Toscano (2009: 396) compellingly argues, Gilbert Simondon provided Deleuze and Guattari with a means of presenting class antagonisms as distinct decoded flows (of money-capital and labour) that form an irresolute social body. As Guchet (2012: 83) clarifies, part of Simondon's response to cybernetics entailed a redefinition of individuation as such a *process* involving human, nonhumans, and machines. His approach exhibits a »refusal to cut off the human from life in general« and begins with processes of partial individuations of species so that »the individuation of the living being and the constitution of its milieu are contemporary and complementary«. Inter-species relations exist (like labour) apart from capital flows and the fiction of unity with nature. Based on racialised myths, this bifurcation inhibits clear, active de-/re-territorialisations of human-nature relations. The Japan-nature fiction continues to subjugate lands under the guise of nationalist impulses, yet with increasing responsiveness to at least one species, the oriental stork. This human-stork interaction asserts an entangled individuation (in Simondon's sense) that may seem to some to support Japan-nature while to others represent what is in effect a posthuman achievement, a new bar for inter-species co-operation within a process of becoming visible in light of the tilt toward biocentrism in Deleuze and Guattari's (1987) approaches to territoriality. What united Japan, the constructed territorialisation of the Japanese archipelago as linked with a unified Japanese race, was also linked to a postwar demilitarised offspring of nationalism in the form of a nature-Japan, which offers the added touch of modelling an implied innocence rather than the countenance of a just-mobilised nation.[15]

Moreover, with the rise of the individualist neoliberal model, citizens in Japan are pitted against corporate interests and more perniciously corporate *models* of how force-relations are mediated and resolved. Local governments themselves must exploit revenue opportunities, so that the stork-reintroduction program becomes a tourist attraction. Civic figures involved in the stork reintroduction project publicly worry over maintaining the local edge in stork-reintroduction leadership in Japan, not to be outshone by another wealthier or savvier prefecture or city in Japan also interested in marketing stork tourism and brands.[16] This attitude reveals how humans are already coveting control over areas not primarily in the interest of the rewilding of storks but rather in initiatives that are founded in competitive inter-human territorialisations of wetlands and paddies.

The connection between the ideological supporting discourses in Japanese state-formation and engagement with international capital flows—rarely dwelled upon in ecocriticism—is habitually self-censored from most documents and continues to blind inter-species understanding on the part of humans. What thus needs to be highlighted is that an effective alliance between nationalist nature-Japan advocates and the neoliberal hegemony suggests that, barring a monetary profit margin, it is difficult to see how pragmatic changes resulting from conservation efforts could take place *en masse*.

—Dean Anthony Brink, National Chiao Tung University, Hsinchu, Taiwan

References

Asquith, Pamela and Kalland, Arne, Japanese Images of Nature: Cultural Perspectives, Richmond, Surrey, 1997.
Brink, Dean, »Review of Anne Allison«, *Precarious Japan. Sungkyun Journal of East Asian Studies* Vol. 14, No. 2, 2014), pp. 279–282.
Clark, Timothy, *Ecocriticism on the Edge: The Anthropocene as a Threshold Concept*, London New York, 2015.
Colebrook, Claire, *Essays on Extinction*, Ann Arbor, 2014.
Deleuze, Gilles and Guattari, Félix, *A Thousand Plateaus: Capitalism and Schizophrenia*, Minneapolis, 1987.
Guattari, Félix, *The Three Ecologies*, London New Brunswick, NJ, 2000.
Guattari, Félix, Genosko, Gary & Hetrick, Jay, *Machinic Eros: Writings on Japan*. Minneapolis, 2015.
Guattari, Félix, and Nadaud, Stéphane, *The Anti-Oedipus Papers*, New York, 2006.
Guchet, X. (2012). Technology, Sociology, Humanism: Simondon and the Problem of the Human Sciences. *SubStance, 41*(3), 2012, pp. 76–92.
Hayles, N. Katherine, *How We Became Posthuman: Virtual Bodies in Cybernetics, Literature, and Informatics*, Chicago, 1999.
Investigative Committee on the Reintroduction of Oriental White Storks, *Assessment of the Extension of Activities for the Reintroduction of Oriental White Storks*, Cooperating Partners of the Investigation Project on the Reintroduction of Oriental White Storks, June 2014.
Iwatsuki, Kunio et al., Sustainable Use of Biodiversity, with Reference to the Japanese Spirit of Worshipping Nature, in: *Conserving Nature: A Japanese Perspective*, Biodiversity Network Japan, 2008, (4–11), Retrieved from https://www.cbd.int/doc/external/cop-09/bnj-nature-en.pdf. Accessed 4 February 2017.
Lorimer, Jamie, *Wildlife in the Anthropocene: Conservation after Nature*, Minneapolis, 2015.
Marks, John, »Information and Resistance: Deleuze, the Virtual and Cybernetics«, in Buchanan, Ian and Parr, Adrian, *Deleuze and the Contemporary World*, Edinburgh, 2006, pp. 194–213.
McCurry, Justin, Japanese fishermen begin annual slaughter of hundreds of dolphins, *The Guardian*, 21 January, 2014, Retrieved from https://www.theguardian.com/world/2014/jan/21/japanese-fishermen-begin-annual-slaughter-of-hundreds-of-dolphins
McRae, James, »Triple-Negation: Watsuji Tetsuro on the Sustainability of Ecosystems, Economies, and International Peace«, in: Callicott, J. Baird, McRae, James, *Environmental Philosophy in Asian traditions of Thought*, Albany, 2014, pp. 359–75.
Moore, Jason W., *Anthropocene or Capitalocene?: Nature, History, and the Crisis of Capitalism*, Oakland, 2016.
Murata, K., Satou, M., Matsushima, K., Satake, S., & Yamamoto, Y., »Retrospective Estimation of Genetic Diversity of an Extinct Oriental White Stork *(Cico-*

nia boyciana) Population in Japan Using Mounted Specimens and Implications for Reintroduction Programs«, *Conservation Genetics* 5, 2004, pp. 553–560.

Ohsako, Yoshito, »Reintroduction Project of the Oriental White Stork for Coexistence with Humans in Satoyama areas, Hyogo, Japan«, Satoyama Initiative, 2011, Retrieved from http://satoyama-initiative.org/reintroduction-project-of-the-oriental-white-stork-for-coexistence-with-humans-in-satoyama-areas-hyogo-japan/. Accessed 10 October 2016.

Ootagaki, H. »Reintroduction of Oriental White Storks in the Wild as the Central Focus of Policy Making«, in: *Conserving Nature: A Japanese Perspective*, Biodiversity Network Japan, 2008, pp. 36–41, Retrieved from https://www.cbd.int/doc/external/cop-09/bnj-nature-en.pdf. Accessed 4 February 2017.

Pickering, Andrew, (2010). *The Cybernetic Brain: Sketches of Another Future*, Chicago, 2010.

Reitan, Richard, »Ecology and Japanese History: Reactionary Environmentalism's Troubled Relationship with the Past«, *Asia-Pacific Journal: Japan Focus*, Vol. 15, No. 2, February 1, 2017. Retrieved February 1, 2017.

Schwägerl, Christian, *The Anthropocene: The Human Era and How It Shapes Our Planet*, Santa Fe, 2014.

Simondon, Gilbert, »The Genesis of the Individual«, in: *Incorporations*, ed. Crary Jonathan, and Kwinter, Sanford, New York, 1992, pp. 297–319.

Stewart, Kathleen, *Ordinary Affects*, Durham, NC, 2007.

Thomas, John, *Reconfiguring Modernity: Concepts of Nature in Japanese Political Ideology*, Berkeley, 2001.

Thrift, Nigel, *Non-representational Theory: Space, Politics, Affect*, Milton Park, Abingdon, Oxon New York, 2008.

Toscano, Alberto, »Gilbert Simondon«, in: *Deleuze's Philosophical Lineage* Jones, ed. Graham and Roffe, Jon, Edinburgh, 2009, pp. 380–398.

Notes

[1] The list offered in the assessment project includes: The Agency for Cultural Affairs; Ministry of Land, Infrastructure, Transport and Tourism; Ministry of Agriculture, Forestry and Fisheries, Ministry of the Environment, Hyogo Prefecture and Toyooka City.

[2] The missteps in building and managing the Fukushima nuclear reactors, including, in brief, building reactors on an underground stream; not heeding warnings predicting exactly what happened five years before the disaster; not evacuating affected areas immediately as public relations trumped living beings in the area, all are well-documented and common knowledge.

[3] See Brink (2014).

[4] To be clear, posthumanism is not related in my ken of use with transhumanism, which focuses on improvements to the human by way of genetic and prosthetic upgrades. However, aspects of the posthuman in the sense of no longer humanistic or focused on human cultures may occasionally overlaps with issues raised (if only by way of a deconstruction of related assumptions).

[5] Colebrook (2014: 20) writes that »all the posthuman celebrations that there is no such thing as ›man‹ and that we are really always already at one with one web of life,

we might ask how it is possible for humans to have this panicked (or joyous) apprehension of self-loss«.

[6] Moore (2016: 6) writes: »the Capitalocene signifies capitalism as a way of organizing nature—as a multispecies, situated, capitalist world-ecology« and »and the era of capitalism as a world-ecology of power, capital, and nature«.

[7] See Simondon (1992: 300) on »the process of individuation rather than the process of individuation by means of the [preconstituted] individual«.

[8] Although John Marks (2006) argues that cybernetic models are too limiting for engaging materiality in the senses Deleuze and Guattari explore, he focuses on human interaction in complex systems that nevertheless share human expectations, whereas inter-species territoriality may indeed benefit from cybernetic metaphors and models of mutually unknowable complexity.

[9] One of most compelling engagements with the cybernetics in Deleuze and Guattari is found in Toscano (2009).

[10] The Satoyama Initiative, or the International Partnership for the Satoyama Initiative (IPSI), is an ecological model based on the promotion of productive human and nonhuman species cohabitation and uses of land. In Japan, it invokes an image of mountain hamlets with rice fields, whereas in other countries it simply suggests rethinking human-environment configuration along less exploitative and more co-productive models.

[11] If may be helpful to point out that since the Meiji period (1867–1912) official and unofficial formulations for the coexistence of west and Japan included slogans such as ›Japanese spirit, western technology‹ *(Wakon-yōsai)* that still leave their mark on commonsense in Japan. Prominent offshoots to appear in cycles later include ›returning to Japan‹ *(Nihon e no kaiki)*, which martialled broad modernist attempts to recover some form of residual Japanese culture and ethics lost during modernization, precisely at the time approaching and during the Pacific War. To complicate matters, the pattern for such nationalist formulations began in the seventeenth century, when Japanese turned to ancient Japanese texts to rediscover a lost Japanese culture before the putative deleterious influence of *Chinese* writing and culture.

[12] See J. McRae (2014).

[13] For an argument exposing fascist elements in ecological thinking today in Japan, see Reitan (2017).

[14] See Murata, Satou, Matsushima, Satake and Yamamoto (2004: 554).

[15] Note the wording in an abstract of a paper cited above: »The development of the Japanese Archipelago was followed by the concept of harmonious co-existence between nature and mankind. However, this traditional concept has now nearly been forgotten even by the Japanese themselves. To establish a sustainable use of resources, such a concept should be understood more widely and its underlying idea should be remembered globally« (Iwatsuki 2008: 4).

[16] The Investigative Committee on the Reintroduction of Oriental White Storks (2014: 27) writes: »With the expansion of activities across the country, there could be concern that the originality and leadership of the activities in the Toyooka region may become overlooked. As a region with a history in relation to Oriental white storks, the parties in relation to the Toyooka region must continue to carry out activities towards higher goals and make efforts to remain the leading communities involved, suggesting a human pride that would place civic preeminence over the welfare of the storks.